Unmasking Safety's Myth of Individualism

Unmasking Safety's Myth of Individualism identifies a critical gap in current safety thinking: safety is not solely an individual responsibility but a social outcome that entails collective action and thinking. Focusing on individual responsibility often translates into individual blame (frequently disguised as accountability) and is usually accompanied by a reluctance to recognize the psychosocial harm that leads to physical and emotional injuries. The cost of this oversimplification is that we lose the vital role of psychological well-being and social dynamics.

This book encourages us not to dismiss the importance of safety management systems, but to expose their limitations. Along with practical pointers, it advocates for a relationship-centered evolution that prioritizes human connection to break through the current status quo and reduce serious injuries and fatalities. The focus on relationship-centered leadership and the critique of an exclusively individualistic or physical focus is timely and relevant. The book challenges leaders to rethink their approach to safety, moving beyond over-reliance on individual responsibility and embracing a relationship-centered model that fosters collective action and psychological well-being. It provides a blueprint where leaders act as catalysts for change, guiding and supporting their teams to achieve shared safety goals.

Unlike traditional approaches, this book explores the crucial role of social dynamics and relational leadership in creating truly safe workplaces. It offers specific strategies for fostering communication and building trust. Additionally, this book provides valuable insights into the role of occupational health and safety (OHS) advisors from a social perspective. It explores how leaders can effectively leverage the health and safety function by understanding the challenges OHS professionals face in influencing organizational culture. Finally, it offers practical advice and strategies for executives and managers as well as for OHS advisors on navigating hierarchical structures and effectively driving safety improvements.

This book is written for all safety leaders, executives, and safety professionals. It could also have a role as supplementary reading for graduate programs.

Rosa Antonia Carrillo, President of Carrillo and Associates, is an internationally recognized leader in creating transformational change with a focus on the safety, environment, and health function. Since 1995 she has authored ground-breaking articles in the safety field and consulted in multiple industries across many countries on the topic of safety leadership. Edgar H. Schein recommended her book *The Relationship Factor in Safety Leadership* (Routledge, 2019) as required reading for anyone wanting to understand safety culture. In 2023 Rosa published *Health and Safety Leadership Strategy: How Authentically Inclusive Leaders Inspire Employees to Achieve Extraordinary Results*, also with Routledge.

Unmasking Safety's Myth of Individualism

A Leader's Blueprint for Psychological Safety and Organizational Resilience

Rosa Antonia Carrillo

Routledge
Taylor & Francis Group
LONDON AND NEW YORK

Designed cover image: Rosa Antonia 2026

First published 2026
by Routledge
4 Park Square, Milton Park, Abingdon, Oxon OX14 4RN

and by Routledge
605 Third Avenue, New York, NY 10158

Routledge is an imprint of the Taylor & Francis Group, an informa business

British Library Cataloguing-in-Publication Data
A catalogue record for this book is available from the British Library

ISBN: 978-1-041-13237-0 (hbk)
ISBN: 978-1-041-13236-3 (pbk)
ISBN: 978-1-003-66877-0 (ebk)

DOI: 10.4324/9781003668770

Typeset in Times New Roman
by Newgen Publishing UK

"Rosa Carrillo continues to challenge the status quo by reminding us that safety is, at its core, a social science. This book is a powerful guide for any leader who wants to move beyond compliance and truly understand that safety is the natural outcome of how we treat, value, and connect with our people. It is essential reading for those ready to lead with authenticity and humanity."

Fernando Carlos, *Head of Safety Culture NFXP Onshore Project, Airswift*

"Rosa confronts a dangerous lie at the heart of modern leadership: that making safety an individual responsibility creates safer organisations. This book names the cost of that lie and gives leaders a practical path to undo the damage it has caused."

Kevin Edwards, *MSc Organisational Behaviour, over 30 years of leadership and organisation development experience, Founder of OB Consulting Limited, UK*

"I have had the honour of working with Rosa on large organisational transformations, where people and trust are the key to sustainable change. Rosa's book offers a vital blueprint for leaders, guiding them away from a flawed emphasis on individual accountability and blame. Instead, it demonstrates how to cultivate a truly resilient organizational culture by harnessing the strength of relationships and collective trust. This book is not a journey map for safety leadership, it is the foundation for a high performing organisation, from which safety naturally flows."

Kelvin Genn, *Chief Disrupter @Art of Work, Co-founder of Safety Differently*

"This is another valuable book from Rosa Carrillo, who is unafraid to challenge the myths of Safety. In this case, the pervasive nature of individualism and, its implications for organisational resilience."

Dr. Robert Long, *Professor, Global Humanistic University, Founder of Human Dymensions*

"This book brings light to the interpersonal dynamics and collective nature of workplace safety, challenging the individualistic narrative. Rosa brings a refreshing approach to helping us to see Safety as a fundamentally relational concept, shaped by how we connect, communicate, and engage with one another in the workplace. For too long, the perception of safety has been confined to engineering solutions and compliance protocols, which overlook the human elements essential to creating a truly safe environment. This book provides us with much needed guidance in the development of the non-technical skills needed to lead in our increasingly complex work environments. It is a must-read for any leader in industry."

Ivan Pupulidy, *PhD, Professor, University of Alabama at Birmingham, Advanced Safety Engineering and Management, Former USFS Director of Innovation and Organizational Learning*

"Rosa's book shows why procedures, engineering, and formal safety systems—while essential—are not sufficient on their own in today's complex, tightly coupled operations. As someone who has written about the limits of rule-based control in high-risk systems, I found her work both extending and sharpening that perspective by showing how relationships, power, and psychological safety ultimately determine whether technical controls succeed or fail, and how organizations must learn inductively by noticing emerging patterns rather than relying only on predefined rules. By emphasizing the quality of everyday conversations as the engine of sensemaking and learning, Rosa offers a practical, human-centered guide to sustaining reliable performance when failure is not an option."

Jim Marinus, *Co-Author of* Critical Steps: Managing What Must Go Right in High-Risk Operations, *Idaho National Laboratory (24 yrs.)*

"People don't get hurt at work because they don't care about safety—they get hurt when systems and cultures make care hard to do. This book makes that distinction impossible to ignore … Safety has never been an individual achievement. What this book reveals is how power, trust, and everyday relationships quietly determine who gets hurt and who goes home safe … Resilient organizations speak sooner, learn earlier, and prevent harm before it occurs. This book explains why relationships, not rules, make that possible."

Dr Andrew Sharman, *CEO & Chair, International Institute of Leadership & Safety Culture, Co-Author of* Mind Your Own Business: What Your MBA Should Have Taught You About Workplace Health & Safety *and* Who Cares Wins: The Psychological Science of Transformational Leadership

Contents

Foreword

For more than 30 years, I have engaged with frontline workers, supervisors, managers, and with executives and directors of boards … advising that safety is not a numbers game, nor is it an engineering problem, or a behavioral compliance exercise. Safety is a *relationship*: a living, shifting, deeply human phenomenon that emerges from how people talk, listen, trust, fear, support, and make sense of their world together.

And yet, safety systems, globally, have been built on an entirely different premise: the myth that if the individual simply tries harder, follows the rules, "takes responsibility," speaks up on command, then the organization will be safe. This myth has dominated boardrooms, audits, investigations, and even so-called leadership programs for decades. It is comforting. It is simple. But it is wrong.

This book dismantles that myth with clarity and courage.

What Rosa has written here resonates with me because it aligns with what I have seen inside hundreds of organizations, across multiple countries, industries, and cultures. Serious incidents are not the result of individuals, workers or leaders, failing to care. They are the result of *systems failing to connect*. When trust collapses, when communication falters, when leaders become distant abstractions, the organization barrels toward a point of failure, of collapse.

In these pages, Rosa reframes safety as a collective, relational achievement, and she does so with a precision that our field has been needing for years. She shows how safety culture is built not by posters and rules but through the daily social fabric of the workplace: how people treat one another, the quality of their conversations, and the beliefs they hold about leadership, power and dignity.

She also highlights a truth that often remains unspoken: the relationship between executives and safety advisors is frequently strained, unappreciated, or misdirected. I have seen brilliant safety practitioners, passionate, competent, insightful, left on the margins of strategic decision-making. It is not because they lack skill, but it is because organizations misunderstand the real nature of their contribution. This book restores that value. It elevates the advisor from a compliance caretaker to a strategic partner in organizational sensemaking.

What makes this work powerful is that it is not ideological, but it is practical. Rosa provides a way forward. She explains how leaders can activate trust, how teams can cultivate shared meaning, and how organizations can move from fragmented awareness to collective intelligence. The Relationship-Centered Safety framework is not another model to memorize; it is a way of seeing and engaging with people that changes how work actually gets done.

The irony is that the safety profession has long insisted on systems thinking while simultaneously clinging to an individualistic worldview. Rosa breaks that contradiction open. She brings us back to the fundamentals: people, emotion, connection, and the invisible threads that hold organizations together, or pull them apart.

Leaders who truly understand safety will find themselves nodding as they read this. Those who have been trapped in entrenched doctrines may feel uncomfortable, and so they should.

This book is timely, necessary, and transformative. It shifts the conversation from blame to understanding, from control to connection, from individual responsibility to shared humanity.

If we can embrace that truth, we will finally begin to build the kind of workplaces we have been promising for decades. Rosa's book sets another great truth in place: we don't create safe workplaces; we create workplaces where safe work can be done and where the people can do safe work.

This book also strongly aligns with my view: safety is *not the absence of accidents,* it is *not the presence of capacity,* but it is the *readiness to respond to risk, relentlessly*. Created and recreated every day, every moment, every job, every project through the interaction of people, systems, and risk.

My hope is simple: executives, safety advisors, and practitioners read this book not as a set of techniques but as an invitation to lead differently, to listen more deeply, and to recognize that safety is something we *create together*, not something we enforce.

The *Unmasking Safety's Myth of Individualism* takes an important step forward in our quest for a safer world.

Corrie Pitzer
CEO, Safemap International
Co-Founder, Safety on the Edge

Preface

On my recent trip to the United Kingdom I heard colleagues complain that the US culture was swallowing their own. The same awakening is occurring in occupational safety and health as countries struggle to implement models designed for the US culture.

Individualism is the dominant story of American culture. It is the narrative that promises us, "You can be anything you want to be." When Americans look back on their history, that appears to be true. Yet, as we look around us in the beginnings of the 21st century, the story no longer matches reality. We see a widening financial gap where a small percentage of people hold most of the wealth, while the rest run faster just to stay in place. Even more unfortunately, this version of corrosive individualism is gaining momentum worldwide.

As a Native American woman in the workplace, I practiced radical self-accountability, taking it to heart that the exclusion and discrimination I experienced would dissipate if I worked harder. Sadly, I was blind to to the realities of racism, insisting that was not limitedd by the rules of social status. I was wrong. Looking back denial was a survival strategy.

Now, awakened I can see that I have a lot to atone for. I spent my career helping companies convince employees that safety is a personal responsibility. I was a genius at enrolling employee engagement. But the evidence was mounting that it was an illusion and I was enabling employers to manipulate people by promising an autonomy that never materialized.

So, here we are. I've recognized that, "You are responsible for safety," is only a slogan. Empowerment is most often a facade to transfer accountability and consequently blame to those lower in the hierarchy. The reality is that eliminating severe injuries and fatalities in the workplace is a collective responsibility. Yet, individuals are the ones punished when something goes wrong. I am not so naive as to say that blame fixes nothing, but I have seen its limitations.

From the beginning of time, humankind survived only by knowing that each member relied completely on the clan. As time passed, that essential truth faded from modern memory, but it remains no less true today. Yes, the clan can also be cruel and ignorant. If we have advanced at all, it should be to recognize when group think needs a correction, rather than disregard the power of relationships and trust. That is precisely where we are now in the workplace. We must remind those who set the price of admission: it is the clan that ensures survival. No one truly succeeds alone.

Now, we are at a crossroads. The fatality numbers are not falling. For a long time, we experienced success by improving physical conditions. Now we are faced with the crux of the problem: power and profit often trump employee well-being. I cringe when I hear the taunt, "Caring is not going to lower fatalities." That phrase is careless and ignorant. We need more caring because powerful people only invest in things they care about. The winning strategy to strengthen organizational resilience and trust is to listen and show that you care by being responsive to people's needs and concerns. Helping leaders and managers see why this is true and how others have done so is my hope for this book.

Acknowledgments

No one writes a book alone, even sequestered; we are always connected. I have people to thank who gave me courage and energy to pursue this topic that feels like a high-stakes undertaking.

First, thank you to Kevin Edwards who spent days reading my work and adding insights that encouraged me to go deeper and see the eight beliefs and relationship-centered safety as a framework to bring us out of the individualistic spell imposed from without and recognize our true generative nature.

Thank you to Corrie Pitzer for his invaluable advice and guidance. I am grateful for his generosity in reading my book and writing the Foreword.

Ivan Pupulidy, thank you for the hours spent editing my book. His efforts made so many things come together in a way that deepened my understanding of social construction and why it matters.

In the same light, I am grateful to one of Ivan's students, Charles Tetteh, who also shared so many insights with me and provided some of the stories that brought the book to life.

Robert Long, you are always a willing listener with a magical power to focus writers and encouraging risky conversations. Our connection is a source of strength and new thought paths to explore.

Thank you to my mother who passed away halfway through the completion of the book. She was the one who always told me I could—especially when others said I could not. I miss you mom but your voice continues within me.

Then there are the women in my life that put up with my ups and downs but never give up on me. We stand together after 40 years of friendship. I see them; they see me, so I exist and we exist together. It would be very hard to write a book without that solid ground we walk together. Kathleen Ramsden, Wendy Williams, Debra Newmark, Herma Rand, Fran Conrad, Ronni Holbrook, Jody Sorkin, Denise Kirwan, Rachel Mickelson, and Jan Matthews, Thank you.

About This Book

Decades of investment in engineering controls and compliance systems have failed to break the plateau of serious injuries and fatalities. The industry hit a ceiling because it treats safety as a technical problem rather than a social science. The "runaway mutation" of individualism—a belief system that isolates workers, obscures systemic failures, and creates a culture of blame severs the connections required for risk detection.

Unmasking Safety's Myth of Individualism is not a denial of individual agency. It is an exposure of how our obsession with individual accountability is a liability. It blinds organizations to the "social relations of work" that produce safety outcomes. By integrating social psychology, complexity theory, and practical case studies, this book reframes risk from an objective number to a socially constructed reality shaped by the presence or absence of trust, language, and power.

For the Executive

Using a social science lens to analyze the Deepwater Horizon disaster, the text illustrates how executives were blinded by the "VIP Visit"—a social ritual that masked the crew's "survival mode" and silenced the weak signals of impending failure. This is contrasted with the Opticore-Tech case, providing a blueprint for evolving from transactional management to "Social Architecture."

For the Occupational Health and Safety Advisor

This book validates the lived experience of the safety professional, moving the role from "compliance officer" to "risk navigator." It acknowledges the emotional labor and burnout inherent in the profession and offers a new identity based on the "Ethic of Care." You will gain strategies to navigate influence without authority, including Non-Technical Skills, the Ladder of Inference, and "Safety Counterinsurgency."

Targeting individual responsibility for safety is a dead end. *Unmasking Safety's Myth of Individualism* challenges leaders to embrace Generativity—the desire to protect future generations—as the ultimate motive for building a resilient, humane, and high-performing organization.

About the Cover

The Mulberry Tree Metaphor

The imagery on this cover is a reflection of the book's central thesis, illustrating that organizational resilience is not the product of individual responsibility but the fruit of deep, systemic interconnectedness. In ancient Chinese philosophical traditions, the symbol of the mulberry tree stands as a defense against stagnation and obstruction. The timeless lesson offered is that when stability is threatened, one must be "bound to a cluster of mulberry trees," a reminder that true strength is found by anchoring ourselves in our collective roots and community rather than attempting to stand alone.

The gnarled, prominent roots reaching into the earth represent digging deep to access the invisible web of relationships that thrive beneath the surface of any successful organization. Within the collaborative circle, the figures gathered beneath the canopy represent a diverse circle of individuals—from executive leadership to front-line workers—engaged in active dialogue.

The art visualizes the results of following a "Blueprint" for a culture where the ethic of care and generativity motivate every person to contribute their talents. Finally, the abstracted urban landscape in the background serves as a reminder that while modern work unfolds in a world of steel and glass, our safety and resilience remain rooted in our most fundamental human instinct: the need for belonging and caring for each other.

This cover was personally illustrated and designed by the author to serve as a visual extension of the leadership principles found within these pages.

1 Embracing Safety as a Social Science

Introduction

Why is Individualism a myth? Why does it need unmasking? Before pursuing these questions, let's clarify. A myth is a cultural artifact that gives meaning to a culture; it is not good or bad. It is a foundational story that explains the world and holds a society together.

My intention, therefore, is not to abolish the myth but to bring it into the open. Failing to scrutinize this complex situation allows an unconscious collection of assumptions to persist, ultimately leading to mistaken decisions and actions to improve workplace safety.

In safety the myth takes form as "I am responsible for safety." When we view safety primarily through the lens of individualism, it highlights individuals making personal choices. It ignores the coercive influence of peer and production pressures. This book unmasks the dark side of this myth and reveals why focusing on individual responsibility is not the answer to reducing fatalities.

Safety Is a Social Science

Safety is a social science. It's relational. It's time to accept this reality and move safety away from a core value to an outcome shaped by connection, communication, engagement, and competence (Douglas and Wildavsky, 1982; Lumann, 1991; Leveson, 2012; Rasmussen, 1997; Dekker, 2011).

For decades, safety was treated as a technical problem. This relied on engineering, protocols, and compliance, operating on the flawed belief that individual responsibility was the key to preventing failure. Yet workplace fatalities have increased globally (NSC, 2023). Ironically, this deeply individualistic field adopted the concept of safety culture. As a sociological concept, culture cannot be influenced at the individual level; it is a collective set of beliefs and assumptions.

A study with 2,524 soldiers provides some answers, reinforcing that safety outcomes are impacted by the quality of relationships in the workplace where Luria concluded that "It is evident that trust in the leader is related to a high climate level, which in turn is related to better safety performance" (Luria, 2010, p. 1292). People cognitively evaluate the information they are given. If they trust their leader, they

DOI: 10.4324/9781003668770-1

are more likely to believe the information and reach similar conclusions about safety's importance. This helps to build a strong safety climate. "Trust is thus shown to be related to both strong and high climates, both of which are mandatory for safety" (Luria, 2010, p. 1293).

I am one of the authors who unintentionally led the discussion of safety culture astray by advocating for a culture where individuals would speak up and stop unsafe conditions. This became a common definition of a positive safety culture, but it proved to be a trap. Since culture is a collective set of largely unspoken assumptions, beliefs, and norms that guide organizational efforts and outcomes, it cannot be changed solely through individual action.

These flawed beliefs contribute to the persistent high numbers of work fatalities. Safety is a collective outcome, arising from the complex interplay of human interactions, emotions, and cultural dynamics. This viewpoint directly challenges the old slogan, "You are responsible for safety." It is time to expose the damaging excesses of the myth of individualism—a collective belief system that creates the "accountable individual" to shield the underlying power structure from examination.

Thus, the audiences for this book are executives and senior-level occupational health and safety (OHS) advisors. This book is all about safety as an outcome of our workplace relationships. This untapped potential is vast. And, this book exposes a key relationship never discussed but critical to supporting safety: the strategic partnership between the organization's top management and the safety advisor.

I felt encouraged when I came across the book, *Just Right: What Your MBA Should Have Taught You about Creating a World Class Safety Culture*, by Andrew Sharman and Judith Hackitt (2024).

> What's your relationship like with your health and safety practitioner? Is it a true partnership or a devolved responsibility? What do you need to do to increase collaboration and allow him or her to have a recognized and respected voice in the business? How well do you know this person – what are their aspirations?

Having taken part in a two-year Key Executive MBA program, I know that the topic of workplace safety never came up. An effect of this disconnect is the disaffection of an essential safety resource. Extensive interviews with safety practitioners around the world in various industries and positions surfaced that more than half of them were thinking of quitting due to feeling unrecognized and disrespected. They love their occupation but dislike the *toxic culture* of their company. This relationship between OHS advisors and management is a reflection of the greater misunderstanding of the competence and contribution of people working in these roles (Carrillo, 2024).

This view of safety as a collective effort offers consequential insights: systemic failures often stem not from a lack of known solutions but from a failure to learn, communicate, and connect. By exploring safety through the lenses of social psychology, sociology, and organizational theory, we can move beyond a reactive, compliance-driven model. This interdisciplinary approach, drawing from fields like linguistics and cognitive science, doesn't diminish technical aspects but emphasizes social systems, allowing us to address complex, human-centric challenges where traditional safety falls short.

Relationship-Centered Safety (RCS) is a framework that takesthese abstract social concepts and turns them into actionable insights. It has three core elements:

1 Theory Y a philosophical belief in the workforce.
2 Socio-technical systems (STS) as an organizational structure that empowers them.
3 The Eight RCS beliefs as the necessary daily leadership behaviors essential to the emergence of trust and open communication (see Table 1.1).

Both Theory Y and STS were developed and applied by McGregor (1960) during the famously successful transformation at a Proctor and Gamble (P&G) plant.[1] It was also the first example of self-directed teams working without management supervision. This was made possible by management's belief in RCS # 4:people are able and willing to contribute to the success of the enterprise, which is their underlying philosophy and is still reflected in their mission statement. The next section explores eight beliefs about human nature including the P&G belief in people.

RCS and the Eight Beliefs

The RCS strategy transitions organizations from an individualistic to a collective approach to safety. This strategy rests on eight core beliefs (Table 1.1) about human nature. These beliefs were discovered through 30 years of interviewing and observing leaders who built high-performance organizations (Carrillo, 2019).

Leaders who adopt these beliefs are generative leaders who unlock the potential of others through relationships. This encourages both collaboration and the willingness to voice conflicts or new ideas. These beliefs also guide how leaders communicate, find the truth, and make decisions in a way that preserves relationships.

A unique aspect of this approach is that it applies to cultures worldwide. An Health Safety & Environment leader from the Middle East shares: The relationship approach to safety really connects with what I've seen over the years working in the Middle East. A significant part of my work has been around welfare and the social aspects of living and working far from home. When you spend time with the workforce in the camps, in the canteens, during transport or downtime, you realise very quickly that safety isn't just about rules or training, it's about how people feel, who they trust, and whether the environment supports them as human beings or not. When relationships are strong in the teams between workers, supervisors, welfare teams, and leadership, people speak up earlier, look out for each other, and problems surface before they become incidents. When relationships are weak, even the best procedures don't stand a chance. Your point about the social side being

Table 1.1 Eight beliefs of Relationship-Centered leadership

Belief	*Intended Outcome*
1. True communication takes place in the presence of relationship and trust.	To guide behavior that encourages open communication.
2. Inclusion precedes accountability.	To encourage trust, open communication, and innovation.
3. Innovation, resilience, inclusion, and accountability are interdependent.	Collectivism supports innovation, accountability, and resilience.
4. People are able and willing to contribute to the success of the enterprise.	Contributes to a sense of inclusion and belonging. Motivates people to contribute their best efforts.
5. People will speak up to stop an unsafe situation if it is in their interest to do so.	Discover what would motivate people to contribute and reduce barriers to openly speaking of mistakes or errors.
6. Drift is a positive quality of adaptive human behavior.	To acknowledge that drift can be dangeerous but is a natural process by which innovation can emerge.
7. Our pre-judgments and biases can prevent us from finding the truth in what we see and hear.	To create a decision-making process that arrives at the truth.
8. Relationships influence emotions, feelings, and beliefs, which influence decisions.	To understand the role and importance of relationships and acknowledge that feelings and emotions also influence decisions

essential to breaking the stagnation in fatality rates is absolutely true in this region. Welfare, belonging, and dignity should not be an "extras." They are part of safety. (Fernando Carlos, Executive HSE Leader).

These beliefs also provide the enabling conditions for effective compliance and technical adherence. So this approach isn't saying that we should abandon the basics that we know work. It is saying that treating people with respect is the necessary foundation for achieving the compliance that traditionalists value. These beliefs, the definition of an organization as a living system, and why the beliefs are essential to the operation of such an organization will be further introduced throughout the book.

Leaders who act from these beliefs improve all organizational performance when they are supported by organizational structures and systems. To explain why and how we introduce an updated safety systems iceberg with a new name *Socio-Technical Systems Iceberg* (Figure 1.1). Conventional safety management has been fixated on what is visible above the surface metrics and events. This is the accident, the injury, the failure. The reactive response is to "drill down" looking for a "root cause" that too often leads to blaming an individual for an "unsafe act" or "human error"—the very paradigm that has proven insufficient.

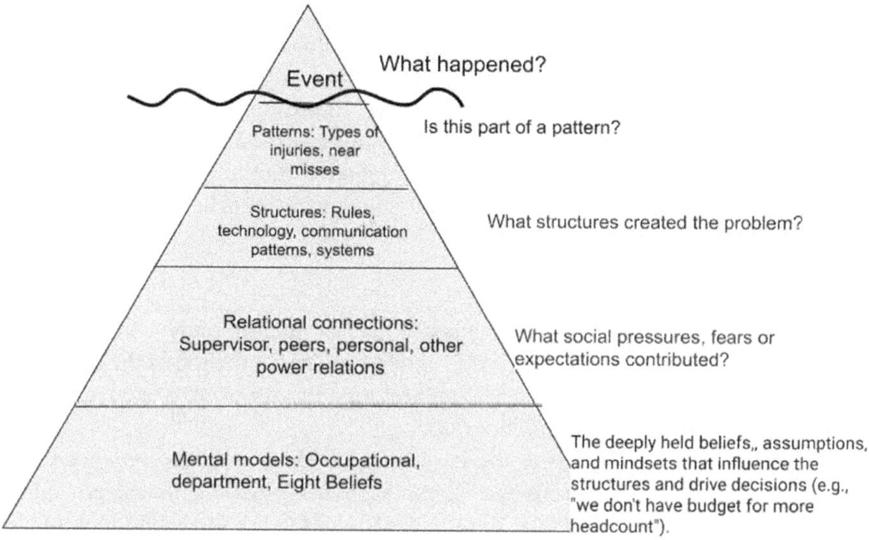

Figure 1.1 Socio-technical systems Iceberg.

The STS Iceberg

The STS Iceberg is a model for prevention and learning from accidents. It is based on the interdependence of social and technical elements. Thus, a change in one level (e.g., structure) will influence how another level operates, affecting outcomes like injuries.

The STS Iceberg serves two functions: it is a forensic tool for deconstructing failures after they occur and, more importantly, a proactive design tool for building a resilient culture. While we introduce the complete structure here, the true complexity and interplay of each layer will only be fully understood after exploring the concepts of social construction, language, and leadership in detail. The goal is to use this iceberg less as an autopsy tool and more as a blueprint for maintaining a healthy, thriving organization. The iceberg will be more fully explained in Chapter 5 with a case study.

Blueprint for the Leader's Role in the Social Systems

The STS Iceberg model guides a leader to shift their identity from a technical manager to a social architect. Instead of focusing only on the visible "Event" at the tip, a leader's new role is to become a social architect whose primary responsibility is to build the vital network of relationships, communication pathways, and shared norms that truly determines an organization's health and resilience. This new role involves planting and cultivating the eight beliefs into mental models. In doing so, the leader sets the "Path of least resistance" toward preventing negative events. This phrase was coined by sociologist Allan G. Johnson (2018) to describe how people often follow the norms of society without question because it is easiest. Of course this path can go in a positive or negative direction. Positive when it contributes toward building a positive society and negative when it is destructive.

Application of STS Iceberg

The iceberg provides a map and methodology for intentionally constructing a relationship-centered culture. Most safety strategies start at the top of the iceberg—trying to "manage" the event. The STS model flips the script. As social architects we begin building from the ocean floor up.

The strategic design begins at the base of the iceberg with the organization's deepest beliefs and assumptions that form the mental models. These are typically passed on unconsciously. The Eight Beliefs of RCS are the generative mental models that a leader must examine and cultivate to create a resilient organization.

A leader begins by examining his/her own beliefs about human nature at the lowest level of the iceberg. Do they believe "People are able and willing to contribute to the success of the enterprise" (Belief #4) or "True communication takes place in the presence of relationship and trust" (Belief #1)"? If yes, that will lead to treating people with respect and communicating effectively. Thus, they begin to build the relational layer. These powerful beliefs are predictive of organizational performance because people contribute in equal measure to how much they are valued and respected.

The relational layer is a unique layer that makes RCS work. The positive or negative aspect of this layer in an organization is a direct result of the leadership mental models. This is most important because the quality of relationships is a foundational determinant for the success or failure in an organization. The relational layer must be intentionally built as a social net, directly countering the isolation, fear, and silence created by the individualistic culture. It is built by systematically cultivating trust, open communication, and psychological safety.

The relational layer is the crucial link that makes the entire iceberg model function. It translates the abstract mental models at the base into the tangible, positive patterns at the top. Without it, even the best-designed structures will fail. A "Stop Work Authority" policy is just a piece of paper in a culture of low trust. But in a relational culture it becomes a vital, collaborative tool rooted in team solidarity.

It is a network of connections that allows information to flow freely and honestly. Without this layer, communication is just a transaction—orders are given, and reports are filed. This layer is where psychological safety is created or destroyed.

It's not a top-down policy but a social achievement that emerges from the daily actions and reactions of everyone involved.

Now begins the construction of *the path of least resistance* toward the prevention of industrial accidents. These are not primarily caused by individual "unsafe acts" or "unsafe conditions," but are instead produced by the "social relations of work" (Dwyer and Raftery, 1991). They are defined as the way the relationship between people and their work is managed within an organization. There are three distinct levels: (1) What is rewarded? (2) The control of how work is done (designed/managed). (3) How power is used and distributed. The *soft infrastructure* is only as good as its supporting structures.

When these layers align, healthy patterns emerge naturally. You don't have to force people to speak up; they do it because it feels safe. You don't have to mandate collaboration; it happens because the relationships are strong. Finally, the Event—the accident—is no longer a surprise to be managed but a rare failure in a resilient system.

The event at the top is the most visible, measurable consequence.

The Iceberg isn't just a diagram; it is an organizing principle. It explains why the technical focus fails and justifies the hard work of building the "soft" layers. Without this map, leaders are navigating the deep waters of organizational culture blind.

Generativity: Motivation for Leaders to Adopt a Social Systems View of Safety

Generativity is the bridge from an exclusively individualistic mindset to a collective, human-centered one. Generativity is the motivation to guide and support younger generations, leaving a positive legacy (Erikson, 1982; McAdams & de St. Aubin, 1992). In the workplace, generativity is a counterpoint to a destructive individualistic mindset that emphasizes self-preservation and often leads to negative outcomes like isolation and burnout. Instead, generativity provides a motivational "why" for a leader: Why embrace collective responsibility and foster an "ethic of care" essential for organizational health? Generativity provides the scientific and psychological foundation for a human-centered approach to safety, providing the intrinsic motivation for building strong, supportive workplace relationships (Doerwald et al., 2020).

Essential Non-Technical Skills

This book provides tools for the generative leader. First among them are the non-technical skills (NTS). First explored in the search to prevent aviation accidents, they found compelling evidence that failures in major disasters were not due to a lack of technical expertise but rather deficiencies in what came to be called NTS. Often dismissed as soft skills, NTS are the key to solving the primary barrier posed by an individualistic culture—the breakdown of communication due to lack of trust and fear of losing face or repercussions. A generative leader is highly competent in these skills and Chapter 7 will cover why and how they are so important in depth.

Leadership Language Shapes Risk Perception

One of the signs of weak NTS is the linguistic divide between workers and management. Executives often use abstract terms like safety culture and risk appetite. But middle managers and employees speak a more tangible language of "burnout" and "excessive workload." This language gap hinders shared understanding and collaboration, which are central to a collective view of risk and how to manage it. Later in Chapter 6, the Mann Gulch disaster case study will show what happens when life-saving knowledge cannot be transferred from those who have it to those who need it most.These problems are not easily solved, especially by orthodox safety methodology because safety is a product of interactions between people and with technology. Whenever people are involved, the organization becomes a living system. This allows for the metaphor of developing an immune system that detects danger and prevents harm.

OHS as an Organizational Immune System

An organization, functioning as a living system, requires an "immune system" for sustained health. This is supported by a "soft infrastructure" that encompasses the essential mechanisms and characteristics of a healthy organizational immune system. It cultivates employees as *risk sensors,* enabling them to detect subtle, intangible signs of potential hazards, similar to how a body's immune system identifies low-grade fevers. When elements like trust, collaboration, and communication pathways are recognized as core strategic assets, on par with financial or physical capital, the organizational soft infrastructure serves as a safety defense system. This approach makes the defense system significantly more efficient and effective.

Acknowledged Limitations

The success of this approach depends on the quality of an organization's relationships. These are actively shaped by those in power. While the methodology is highly transferable, a dedicated leadership team or a single team leader is required to initiate the work to build these working relationships.

Changes in language and conversations alone do not produce results. Research shows that a manager's demonstrated commitment to safety is the most effective predictor of success. What does that commitment look like? There is strong empirical evidence that a senior manager's commitment to safety is demonstrated through social and problem-solving skills, as well as technical safety knowledge (Fruhen et al., 2014).

While these internal leadership skills are vital, for many large corporations, significant change will not occur without external societal pressure. An analysis of landmark workplace disasters shows that while industrial contexts have changed, the power dynamics leading to tragedy remain remarkably consistent. These disasters are not isolated failures but a persistent, individualistic pattern of prioritizing production and profit over worker safety (Fruhen et al., 2014). In *The*

Relationship Factor in Safety Leadership (2019), I provide examples of leaders who stepped off the negative path of least resistance and how they did it. After years of teaching this material, I have come to realize that their approach is simple but not easy for most people to employ.

Conclusion

This chapter challenged the industry's oldest assumption that safety is an individual responsibility. We have established that safety is a social science, an outcome of relationships rather than just engineering. As the global fatality rates indicate, the technical approach has hit a ceiling; we must now look deeper.

To do this, we introduced the STS Iceberg. While traditional safety obsesses over the visible "Event" at the tip, true prevention happens in the submerged layers of Relationships and Mental Models.

Finally, we identified the fuel for this transformation: Generativity. The desire to leave a legacy is the only motivation powerful enough to replace the fear-based compliance of the past. The Eight Beliefs introduced here are the tools to operationalize that shift, moving us from the seductive simplicity of blame to the complex, effective work of human connection.

Note

1 P&G plants that utilized Theory Y and STS suppressed production and profitability so far above the others that they were accused of faking the numbers. They proved true when audited.

References

Carrillo, R.A. (2019). *The Relationship Factor in Safety Leadership*. Routledge.

Carrillo, R.A. (2024). *OHS Voices from the Resistance*. Self-published.

Dekker, S. (2011). *Drift into Failure: From Hunting Broken Components to Understanding Complex Systems*. Ashgate.

Doerwald, F., Zacher, H., Van Yperen, N. W., & Sheibe, S. (2020). Generativity at work: A meta-analysis. *Science Direct*. www.sciencedirect.com/science/article/pii/S0001879120301469

Douglass, M., & Wildavsky, A. (1982). *Risk and Culture*. University of California press

Dwyer, T., & Raferty, A.E. (1991). Industrial accidents are produced by social relations of work: A sociological theory of industrial accidents. *Applied Ergonomics*, 22(3): 167–178.https://doi.org/10.1016/0003-6870(91)90156-C

Erikson, E.H. (1982). *The Life Cycle Completed*. W. W. Norton & Company.

Fruhen, L., Mearns, K., Flin, R., & Kirwan, B. (2014). Skills, knowledge and senior managers' demonstrations of safety commitment. *Safety Science*,69: 29–36.

Johnson, A. G. (2018). *Privilege, Power and Difference*. McGraw-Hill Education.

Leveson, N. (2012). *Engineering a Safer World: Systems Thinking Applied to Safety*. MIT Press.

Luhmann, N. (1991). *Sociology of Risk*. De Gruyter.

Luria, G. (2010). The social aspects of safety management: Trust and safety climate, *Accident Analysis & Prevention*, 42(4): 1288–1295.

McAdams, D. P., &de St. Aubin, E. (1992). A theory of generativity and its assessment through self-report, behavioral acts, and narrative themes in autobiography. *Journal of Personality and Social Psychology*, 62: 1003–1015.

McGregor, D. (1960). *The Human Side of Enterprise*.McGraw-Hill.

NSC (2023). *Workplace Deaths up 12% Worldwide in the Past Two Decades: Report.* National Safety Congress. www.safetyandhealthmagazine.com/articles/24808-workplace-deaths-up-12-worldwide-in-past-two-decades-report

Rasmussen, J. (1997). Risk management in a dynamic society: A modelling problem. *Safety Science*, 27(2–3): 183–213.

Sharman, A., & Hackitt, J. (2024). *Mind your business: What Your MBA Should Have Taught You about Creating a World Class Leadership Culture*. Maverick Eagle Press. https://archive.org/details/mindyourownbusin0000shar

2 The Problem with Individualism in Safety

Introduction

Fatalities in the workplace represent a profound failure of modern societal values and an unacceptable cost in a developed world. There are instances where people risk their life with full knowledge of the possible consequences. Firefighters and police come to mind, and we are grateful to them.

The very idea that individuals face fatal risks simply by performing their daily work tasks is a stark contradiction to the advancements in safety, technology, and human rights that define our contemporary world. As a "hangover of the organization as a machine metaphor" (Itkin & Nagy, 2014), this situation demands a re-evaluation of current safety protocols, accountability, and the underlying cultural attitudes, such as that individuals are responsible for safety in the workplace.

Chapter one provided a broad view of safety as a social science. This chapter provides in-depth inquiry into why current approaches to lowering fatalities are inadequate by directly challenging the long-standing belief that workplace safety is primarily an individual responsibility. It highlights the threats present in this individualistic ideology, noting that despite significant investments, the United States, which ranks highest on the cultural individualism scale, still has the highest fatality rates in the developed world. Chapter 3 goes more deeply into the historical development of individualism.

Individualism as a Global Concern

A recent global analysis published in *Psychological Science* (Santos et al., 2017) shows that almost all countries—especially those exposed to US media and economic models—are drifting toward greater individualism. This includes Asia, Western Europe, and other close partners of the United States.

As the researchers explain,

> The fact that most of the countries that did not show an increase in individualist values were among the lowest in socioeconomic development over the time period examined is consistent with the observation that socioeconomic development drove the rise in individualism [...] China is an exception to this

DOI: 10.4324/9781003668770-2

> pattern, showing a decrease in individualist values even though the country has experienced economic growth. Notably, China has a complex socioeconomic history, so it will be worthwhile to investigate this country in more detail in future research.
>
> (Santos, et. al, 2017)

The perils of excessive individualism are not limited to the United States. Other countries are adopting its negative characteristics. It extends globally and is being researched as a cause for a rise in fatalities in Asia (ILO, 2023). This sets the stage for arguing that an emphasis on individualism is not suited for achieving workplace safety.

OHS provides a canvas to expose the manipulative nature of what has long passed for employee empowerment in organizations. Current safety discussions, particularly on social media, gravitate toward a familiar set of pillars: robust engineering controls, psychological safety, individual accountability, exemplary leadership, a resilient safety culture, and employee engagement. These concepts are a staple of conferences, webinars, and policy documents. Also on this list are the typical topics of process safety management, risk management/control, and safety management systems. All of which leave out the social relational aspects of work and the acknowledgement that there is no zero risk option in what we humans do. Instead the industry has socially constructed the belief that we can control complex systems (Long, 2019).

Yet, industrial disasters continue despite widespread discussions on these interventions. We are forced to ask a pressing question: Are these concepts of risk management truly meeting organizational needs? While one might argue that perfect application of those fundamentals would prevent disasters, this perspective overlooks a significant dimension. The failures we observe are not merely deficiencies in the application of known remedies but a deeper systemic issue rooted in the overlooked complexities of human interaction within dynamic, often ambiguous, operational contexts (Flin et al., 2008).

Unless we subscribe to the notion that every disaster listed in Exhibit 2.1 is a direct consequence of conscious negligence by executives, managers, or workers, we are compelled to consider deeper, intangible explanations. This book proposes answers from the field of social construction, arguing that the very meaning of safety, risk, and even our understanding of "facts" is not an objective truth. Risk is collaboratively defined through human interaction and agreement (Gergen, 2009). This continuous, collective process, known as sensemaking, shapes how individuals and groups perceive, interpret, and respond to surprise and uncertainty.

A singular focus on individual accountability often obscures the collective, socially constructed realities that truly drive risk perception and workplace outcomes. By understanding these deeper, intangible forces, we can move beyond superficial solutions and recognize the systemic vulnerabilities that contribute to tragic workplace fatalities.

Cultural Influences

There is some evidence that a country's culture influences the fatality rate. Significantly, the US fatality rate, at 3.5 per 100,000, is double that of the EU. It is notable that the United States holds the highest fatality rates and scores highest on Hofstede's Individualism scale, with 91 out of 100 points (YouFactor, 2024). This complexity helps to explain why imported safety programs often fail and why a one-size-fits-all approach is not effective. However, a direct, singular link to workplace fatality rates is complex and often mediated by a multitude of other factors, such as the strength of safety regulations, industry type, and enforcement mechanisms (Bush et al., 2021; Chen, 2019; Choi, 2010).

For example, when comparing the United Kingdom to the United States, the United Kingdom, with a fatality rate of 0.42 per 100,000, has historically had more government oversight and spent more on safety. Unfortunately, increased spending and government oversight have shown little impact on US fatality rates. More spending is not a silver bullet. Like all safety approaches, if the culture is not prepared to accept them, they will not take root.

The United States' comparatively weaker workplace safety record is a complex issue with no single cause. However, its brand of particularly strong individualism likely contributes to the resistance against stricter regulations, increased risk-taking behavior, and deviation from standard procedures. Quantifiable research during the COVID-19 pandemic showed that collective cultures (Japan, China) were more likely to comply with public health interventions like mask-wearing and social distancing, even when it meant personal inconvenience (Leong et al., 2022; Lu et al. 2021; Rajkumar, 2021).

The Limitations of Individualism in Workplace Safety

In an individualistic culture, the whole person is often diminished. The individual, as a physical entity, can be identified and managed, while intangible factors like emotions and memory are dismissed because they are not easily quantified or controlled. As a result, workers are considered physical assets, but not quite as important as the structural assets. This became clear to me when a high-level manager at a mine had the insight that he knew all about his physical assets and that he knew very little about the people working in the mine (Carrillo, 2019).

In workplace safety, individual responsibility is often a deceptive oversimplification. While intended to foster accountability, it frequently devolves into individual blame (Chikudate, 2009).

Safety Culture: A Collective Concept Turned Individualistic

It is ironic that the OHS industry turned to the collective concept of safety culture to lower accidents. Culture is created and sustained by groups of people. An individual cannot be a culture.Thus, an industry steeped in individualism transformed safety culture into the ideal of having every employee willing to take on the

Exhibit 2.1 Examples of Industrial Disasters

August 7, 1994: Australia. Eleven workers lost their lives in an explosion in the Moura No 2 underground mine.

April 5, 2010: Upper Big Branch Mine disaster, West Virginia, United States. An explosion occurred in Massey Energy's Upper Big Branch coal mine. Twenty-nine out of 31 miners at the site were killed.

July 6, 2013: Lac-Mégantic, Quebec, Canada. Lac-Mégantic derailment. Forty-seven people were killed when there was a derailment of an oil shipment train. The oil shipment caught fire and exploded, destroying more than 30 buildings. It was the fourth-deadliest rail accident in Canadian history.

August 12, 2015: Binhai, Tianjin, China. Two explosions within 30 seconds of each other occurred at a container storage station at the Port of Tianjin in the Binhai New Area of Tianjin, China. 173 people died as a result.

January 25, 2019: Brumadinho dam disaster, Minas Gerais, Brazil. An iron ore tailings dam suffered a catastrophic failure. At least 259 people died.

responsibility to stop unsafe actions and make every effort to follow protocol (Lin, 2012).

An emphasis on collective responsibility, in contrast, can foster a culture of mutual vigilance and support, where individuals actively encourage risk assessment conversations, resulting in a more secure and injury-free work environment. These different cultural approaches profoundly influence whether workplace safety is viewed as an individual or shared responsibility (O'Sullivan, 2018).

Cultural Incompatibilities

Stepping away from radical individualism in safety toward collectivism requires new language and approaches. We can learn a lot from observing safety in countries from collective cultures. One of the most salient characteristics is attention to social interaction like saying "good morning" or asking "how is your family?" Managers who are not aware of such cultural differences often wonder why employees do not share information or bring up problems with them.

The impact of any cultural value on safety is not isolated but is always intertwined with the larger national cultural environment and the specific workplace culture. Research findings, especially those from Western settings, cannot be universally applied across all cultures and industries (EU-OSHA, 2013). A study of organizational, societal culture, and leadership found correlations between national culture and safety incidents. It was based on 62 countries with a questionnaire among 17,000 middle-level managers. An analysis reveals that power distance and individualism versus collectivism are particularly relevant lenses through which to view safety implementation (see Exhibit 2.2). Other empirical research shows that a collective culture provides a strong foundation for a collaborative safety environment. And that a high power distance culture protects against the detrimental side of collectivism, which is a reluctance to challenge a manager or use stop-work authority (EU-OSHA, 2013).

There are advantages and disadvantages to each dimension. This book suggests that OHS—the whole organization—would perform better by adopting the following positive values of collectivism (see Exhibit 2.3).

Collective culture characteristics already exist in some companies in individualistic country cultures, indicating that it can be accomplished. . And that is what needs tohappen because a standardized, "one-size-fits-all" safety program is a strategic risk. To ensure effective implementation and genuine buy-in across diverse organizational units or global operations, executives must prioritize cultural intelligence. This means investing in understanding these key cultural dimensions and tailoring safety strategies, communication, and engagement methods accordingly.

Failing to account for these traditional cultural differences will not only undermine the effectiveness of safety programs but also squander resources and potentially increase risk. A culturally informed approach, in contrast, fosters greater employee ownership, enhances communication, and ultimately delivers more sustainable and impactful safety outcomes.

An OHS Practitioner's Defense of Personal Responsibility

Letting go of personal responsibility for safety as the primary defense stirs deep concerns in the OHS practitioner community. The following quote from a practitioner expresses their frustration with new theories while basic safety practices are poorly executed:

> After every fatality, we hear: "This must never happen again.""Safety is now our top priority.""We've implemented comprehensive changes." These statements expose an uncomfortable truth: the solutions were already known; they just weren't acted on until someone died.
>
> When companies are prosecuted, it's never for:
>
> - Lack of human-centered frameworks
> - Insufficient theoretical understanding
> - Inadequate complexity theory integration
>
> It's for failing to:
>
> - Maintain equipment
> - Follow procedures
> - Implement controls
> - Resource safety properly
> - Act on known risks
> - Train workers effectively

Exhibit 2.2 Cultural Factors in Safety Management Design

- In High Power Distance cultures, a top-down, highly structured approach might be more readily accepted, while Low Power Distance cultures thrive on autonomy and worker involvement in decision-making.
- A collective culture is a social structure that prioritizes the needs and goals of the group over the individual's.
- An individualistic culture is a social structure that prioritizes the needs and goals of the individual over those of the group.

Exhibit 2.3 Collectivism Values

- Shared Accountability: Employees feel a strong sense of responsibility for the well-being of their coworkers and the team as a whole. This fosters a "group consciousness" where people look out for each other, which is a key component of a strong safety culture.
- Team-Oriented Communication: Collectivistic cultures often prioritize cooperation and open communication within the group. This can facilitate the quick sharing of safety information and the recognition of hazards, as team members feel comfortable speaking up for the good of the group.
- Peer Influence: When the collective values safety, individuals are more likely to adhere to safety protocols to maintain group harmony and avoid letting their teammates down.

- Enforce existing standards
- Conduct thorough risk assessments
- Provide competent supervision
- Listen to worker concerns

So, my question to you is this: "How do new theories of complexity or emergence offer a better solution than our existing laws, systems, and frameworks? Wouldn't it be more effective to prioritize compliance, accountability, and real-world application instead of expanding theoretical concepts?"

This is an important question, often posed by safety practitioners deeply immersed in the practical aspects of their work. Despite existing laws, systems, and frameworks designed to prevent fatalities, accident investigations consistently reveal a failure to follow established requirements. While companies face fines and executives are held accountable for non-compliance, similar issues regrettably recur in subsequent disasters.

The answer this book provides is that a collective, relationship-centered approach is not a replacement for the basics, but the enabling condition for them. For example, why is stop-work authority so often difficult to implement? It was not because they did not know the policy existed or how to do it. It was because RCS Belief #5, *People will speak up if it is in their best interest*, is not present. The monolith of social pressures prevailed.

A collective, relationship-centered approach does not supersede these foundational practices but rather serves as the enabling condition for them. Through Belief #1, *True communication takes place in the presence of relationship and trust*, open communication is understood and practiced.

Personal Responsibility Re-examined

Personal responsibility emphasizes the individual's autonomy and accountability for their own actions and outcomes, stemming from a belief in individual agency. These beliefs ultimately lead to the finding that human error is a root cause of accidents.

At first, it may feel counter intuitive to say that "individual responsibility for safety" is a mistaken belief. For individualists, it feels like an ethical dilemma to accept they are not 100% in control of what they feel, think, say, and do. However, there can be a substantial gap between reality and how much responsibility individuals are taking. This was demonstrated by surveys at manufacturing sites where 100% of employees responded that they took personal responsibility for safety (Carrillo, unpublished report, 2000). This was puzzling to the OHS department because their OSHA recordable rate was the highest in their industry.

The reality is that individual beliefs and perceptions are powerless against an entrenched group culture. Yet, I can hear people saying, "Individuals can influence profound changes through their curiosity, courage and commitment." I agree that an individual can voice and pursue an idea but unless they gain followers it does not go past being an individual belief.

In fact, some ideas encounter such strong social barriers that no amount of persistence or commitments on the part of an individual can get them accepted. Thankfully, we are past burning people at the stake for refusing to renounce a belief. However, society still has ways to isolate individuals who persist in beliefs that are not culturally approved.

The employees' belief that they took personal responsibility for safety did not make that facility safer. The facility was unsafe because what people *saw and felt* was more powerful than what people *believed*. An individual's good intentions are quickly overridden by the social proof of a group moving in a different direction.

An internal belief that *safety is my responsibility* also must become a group belief. Yet, the concept of control in safety management relies on rote compliance and individual responsibility. Employees are told they must follow procedures. They must report unsafe conditions and near misses. In some companies, they must report unsafe actions they observed among their peers. They must stop an unsafe job and they must offer suggestions to make corrections. Some organizations demand a specified number of reports a period of time. All of this is offered under the guise of empowerment—a misnomer since power rarely accompanies the assigned responsibilities and the response to reports.

Social Relations Rather Than Individual Error

This American-grown philosophy of individualism, reinforced by a mechanistic approach to accident investigation, was exported globally, influencing safety practices worldwide and laying the foundation for decades of safety mismanagement that placed the onus on the individual. This approach is still considered a significant contributing factor to the higher fatality rates in the United States.

This export of individualism created a cultural collision in Asia. In many parts of the continent, rapid industrialization was driven by Western multinational corporations. These companies often brought their entire operational models, including safety philosophies heavily influenced by "Safety First" and Heinrich's individual-focused theories. While many Asian cultures are more collectivist and group-oriented, the imported safety model was based on individual blame.

This led to a compliance-oriented safety culture where workers followed rules to avoid punishment but often lacked the psychological safety to report systemic problems or goal conflicts. The 1984 Bhopal tragedy in India demonstrates this clash of philosophies and the limitations of a purely top-down, individual-focused model. Weick (1988) recounts that in the months before the Bhopal disaster, management demonstrated a collective sense of indifference and a disregard for safety, having decided that the plant was closing and no longer of value. As Weick states, "Top management's perceptions of Bhopal set into motion enactments that confirmed their perceptions. It became a self-fulfilling prophecy" (p. 313). The official toll is 5800 deaths and a possible 20,000 who died of gas related illnesses (Kumar, 2004).

Conclusion

This chapter has provided a critical analysis of how the ideology of individualism has shaped workplace safety, both in the United States and globally. We have seen how telling individuals that they are responsible for safety and blaming them in the investigation process is not lowering fatality rates. The irony is that even as we invest more in safety, we have neglected the very thing that can make those investments work: the natural relationships that bring people together with a common purpose.

We have reached a point where the basics of safety cannot be properly executed without a new, more effective framework. The challenge now is to move from a narrow, individualistic view of safety to one that recognizes its social and relational roots.

The rest of this book will provide that framework. It provides leaders with the tools to cultivate their organization as a living safety defense system, much like the human body's immune system. This is the path to building the trust and communication necessary to make safety an emergent property of the organization, not a separate add-on required by regulators.

References

Bush, C., Usrey, C., Loud, J., & Carrillo, R. A. (2021,). Serious injuries and fatalities: Why are they constant while injury rates decrease? *Professional Safety Journal*, 01: 26–31.

Carrillo, R. A. (1999, 2000, 2024). *Unpublished Safety Assessment Culture Reports.*

Carrillo, R. A. (2019). *The Relationship Factor in Safety Leadership:*. Routledge.

Chen, M. (2019). *National Culture and its Impact on Workplace Health and Safety and Injury Prevention for Employers and Workers*. Superdiversity Institute for Law, Policy and Business. https://papers.ssrn.com/sol3/papers.cfm?abstract_id=394_307

Chikudate, N. (2009). If human errors are assumed as crimes in a safety culture: A lifeworld analysis of a rail crash. *Human Relations*, 62: 1267–1287.

Choi, W. D. (2010). *Structural Model of Effects of Cultural Factors on Escalation of Commitment through Antecedents, Agency, and Negative Framing Effects*. University of Nevada. https://digitalscholarship.unlv.edu/cgi/viewcontent.cgi?article=1761&context=thesesdissertations

Dekker, S. (2011). *Drift into Failure: From Hunting Broken Components to Understanding Complex Systems*. Ashgate.

EU-OSHA. (2013). *Cross Cultural Differences in OHS*. European Agency for Safety and Health at Work. https://oshwiki.osha.europa.eu/en/themes/cross-cultural-difference-osh.

Evotix. (2023). *A Breakdown of HSE's Key Health & Safety Figures for Great Britain (2022/23).* www.evotix.com/resources/blog/a-breakdown-of-hses-key-health-safety-figures-for-great-britain-2022/23

Flin, R., Mearns, K., O'Connor, P., & Bryden, R. (2000). *Safety Climate: AReview of the Literature*. Offshore Safety Division, Health and Safety Executive.

Flin, R., O'conner, P., & Crichton, M. (2008). *Safety at the Sharp End*. Ashgate.

Gergen, K. J. (2009). *An Invitation to Social Construction* (2nd ed.).Sage Publications, Inc.

Hopkins, A. (2005). *Safety, Culture and Risk: The Organisational and Social Foundations of Industrial Safety*. Allen & Unwin.

ILO. (2023) *Nearly 3 million people die of work related accidents and diseases*. www.ilo.org/resource/news/nearly-3-million-people-die-work-related-accidents-and-diseases

International Association of Oil and Gas Producers, IOGP. (2014). *Human factors 101*. https://humanfactors101.com/topics/non-technical-skills-crm/

Itkin, H., & Nagy, M. (2014).Theoretical and practical use of metaphors in organizational development and beyond. *Pannon Management Review*, 3: 37–72. http://pmr.uni-pannon.hu/articles/3_4_itkin_nagy.pdf

John Hopkins University (2023). *Why should mental health be a priority in the workplace?*. https://imagine.jhu.edu/blog/2023/05/17/breaking-the-stigma-why-mental-health-should-be-a-priority-in-the-workplace/?>

Kumar, S. (2004 Aug). Victims of gas leak in Bhopal seek redress on compensation. *BMJ*, 329(7462): 366.

Leong, S., Eom, K., Ishii, K., Aichberger, M. C., Fetz, K., Müller, T. S., Kim, H. S. (2022). Individual costs and community benefits: Collectivism and individuals' compliance with public health interventions. *PLoS ONE*, 17 (11).

Lin, M. L. (2012). *The Effect of Interdependence on Safety Performance and Sustaining Operating Dexterity in a Complex Environment* (unpublished manuscript). Dupont.

Long, R. (2019).*The Social Psychology of Risk Handbook: i-thou*. Scotoma Press.

Lu, J. G., Jin, P., &English, A. S. (2021). Collectivism predicts mask usage during the COVID-19 pandemic. *Proceedings of the National Academy of Sciences*, 118(23).

O'Sullivan, J. (2018). American workplaces are 900 times more deadly than Britain. *ISHN*. www.ishn.com/articles/108262-american-workplaces-are-900-more-deadly-than-british-ones
Rajkumar, R. P. (2021). The relationship between measures of individualism and collectivism and the impact of COVID-19 across nations. *Public Health in Practice*, 2:100143.
Santos, H. C., Varnum, M. E., & Grossmann, I. (2017). Global increases in individualism over the past 50 years. *Psychological Science*, 28(1): 58–71.
Santos, H. C., & Grossman, I. (2017). *Individualistic Practices and Values Increasing around the World*. APS.
Weick, K. E. (1988). Enacted sensemaking in crisis situations. *Journal of Management Studies*, 25(4): 305–317.
YOUFactors. (2024). *How many people die at work each year?* YOUFactors. www.youfactors.com/statistics/how-many-people-die-at-work-each-year-2024-edition

3 Historical Roots and Impact of Individualism

Introduction

While Chapter 2 identified the symptoms of the problem, this chapter examines the pathogen. We are not challenging individualism as a concept, but as a runaway mutation that has infected our safety management systems. To cure it, we must understand how it weakens the very systems designed to protect.

The Myth of Individualism as a Collective Illusion

The truth about individualism is that it is a collective belief system that often serves to obscure and legitimize existing power structures which actively shape the individual into a "subject" that can be monitored and controlled. In a safety context, this means the system creates the "accountable individual" as a convenient way to manage risk at the lowest level, thereby protecting the power structure from scrutiny (Foucault& Gorz, 1989). This ideology, which has been widely exported globally, provides the philosophical justification for placing blame on workers instead of examining the role and responsibility of organizational systems.

In Western culture, this belief system is historically embodied in powerful narratives. Political theorist Alex Zakaras (2022) identifies these myths as a "double-edged sword." They are neutral cultural vessels that can be filled with exclusionary definitions to maintain hierarchies or egalitarian definitions to expand democracy. Whoever controls how this "operating system" is defined controls who gets to participate in society. This is a function of power not truth.

This version of individualism cultivates a workplace culture where fear of ostracism decreases trust and open communication (Flin et al., 2000). Without communication, failure is inevitable. In short, the modern idea of the autonomous individual can paradoxically make people less safe by undermining the social connections and collective structures, and relationships that are essential for increased resilience and well-being (Flin, 1996; Bauman, 2000; Pupulidy and Vesel, 2023).

DOI: 10.4324/9781003668770-3

The Globalization of Individualism

As noted in Chapter 2, there has been a widespread increase in individualistic values, largely influenced by US models. The history of individualism, particularly its alignment with laissez-faire capitalism during the Industrial Revolution, reveals an inherent tension: the pursuit of individual wealth often came at the cost of worker well-being, as companies prioritized profit over safety despite an espoused value of *safety first*. This historical context foreshadows the negative societal impacts when individual interests are not balanced with broader communal concerns.

The prevailing belief that accidents are caused by a worker's "unsafe acts" and "unsafe conditions" is deeply rooted in this individualistic paradigm. This paradigm began with H.W. Heinrich. In 1931, his work *Industrial Accident Prevention* was used to codify the myth that 88% of accidents are caused by human unsafe acts (Bush, 2021). This was an easy fit for a predominantly individualistic culture. Bush argues that Heinrich's insurance data was converted into a "law of nature." Because the United States emerged as the dominant industrial superpower post-WWII, the myth didn't just stay in American factories; it became the global curriculum for safety management. We didn't just export steel and machinery; we exported the belief that human error is the problem in safety, thus, effectively colonizing global safety culture with a uniquely American definition of individual responsibility.

This dangerous oversimplification is directly challenged by Dwyer and Raftery (1991), who found that accidents are produced by the social relations of work—specifically, the systems of rewards, command, and organization. When workers have the freedom and knowledge to control their own work and safety, they are effective at preventing accidents. When they are not effective, management must intervene by focusing on the broader social picture—the relationships, trust, and communication in the workplace. This view suggests that ergonomic and process changes must always consider their interaction with the social relations of work.

The American-grown philosophy of individualism influenced practices worldwide and laid the foundation for decades of safety management that placed the onus on the individual—a philosophy now challenged by a more modern, holistic, and relational understanding of safety.

While the United States often embodies a more competitive and market-driven form of individualism and a distrust of state authority, other highly individualistic Western cultures, such as the United Kingdom, Canada, and Australia, exhibit a gentler form. These nations have a strong sense of egalitarianism and community-mindedness, manifested as communal healthcare funding and investment in public education, which helps to mitigate the more isolating aspects of individualism.

This historical and global trajectory has significantly impacted safety. The prevailing belief that accidents are caused by a worker's "unsafe acts" and "unsafe conditions" stems, in major part, from this individualistic paradigm. They agree that preventing their recurrence requires shifting our attention from the individual to social dynamics within teams and organizations (Flin et al., 2008).

The Paradoxes of Individualism in the Global Workplace

Now we come to the crux of the problem. The global rise of individualism also manifests as a deep and enduring paradox in workplaces worldwide: the rhetoric of personal responsibility and accountability often clashes with the reality of limited employee power and institutional inertia. Employers globally demand that employees take personal responsibility and contribute to the success of the enterprise, yet opposing forces frequently keep people without the power or tools to ensure things go well, thus setting safety and production at odds with each other.

That said, I want to clarify again that individualism is not bad or corrupt. The thing to do is to rediscover and reappropriate some of the more egalitarian iterations of this individualist mythology. We've seen them in the abolitionist movement and the early feminist movement of the nineteenth century. The ideal of personal independence has expressed a very expansive ideal of equality, suggesting that a well-ordered society is one in which individuals own their own property, own their own tools, are not beholden to anyone else, and so control their own economic lives.

Of course this mythology needs to be reformulated to fit the culture of each country and the circumstances of the 21st century economy. But because these myths are so powerful, it's incumbent on those who want to press for greater equality to do what reformers have done throughout history—to reclaim, reinterpret, and redirect that language (Zakaras, 2022).

For example, Nelson Mandela reconciled African collective culture with individual rights by adopting the philosophy of *Ubuntu* —"I am because we are"—integrating it into the 1993 Constitution to prioritize restorative justice over revenge. He balanced these by advocating for the collective liberation of his people while ensuring the new democratic framework legally guaranteed individual freedom and dignity (Unesco, 2024).

The Historical Evolution of Individualism as a Moral Virtue

The current crisis of individualism, marked by anxieties over social isolation and the breakdown of communal bonds, is not a coincidence. It traces its origins back to the Enlightenment, which prioritized reason over tradition. In making this trade—the security of the collective for the liberty of the individual—we felt an initial excitement whose luster has since worn out.

The historical journey toward individualism was fraught with religious conflict. During the Middle Ages, individuals were primarily conscious of themselves as members of a group. The Renaissance marked a shift where man became a spiritual individual, and recognized himself as such. France still retained strong elements of communal identity and ascribed roles, making individual expression potentially dangerous. This historical juncture underscores the complex and often painful processes involved in the societal move from collective identities to the primacy of the individual.

During the early modern period, the orthodox religious beliefs that had always supported society's hierarchical rules began to weaken. This left a gap in how people justified what was morally and ethically right. As a result, the source of authority shifted. Instead of coming from outside, divinely approved rules, or community customs, authority became increasingly centered on the individual. This shift emphasized decentralized decision-making, rooted in subjective processes like reasoning, conscience, and personal interpretation. Consequently, the acceptance of ways to behave depended on the result of each person's internal deliberation. This was a significant change from previous times when right and wrong were most frequently socially determined.

In modern history, individualism has been celebrated as an empowering force for progress, innovation, and national wealth. However, a necessary counter-perspective is that a continued, unchecked emphasis on individualism may be detrimental to societal well-being and, specifically, to workplace safety. The core hypothesis is that individualism, by emphasizing safety as a primary individual responsibility, creates a system prone to blaming people for accidents.

Individualism Today in Safety

The history of safety management is a tale of two philosophies: one born in Europe and the other in the United States. The United Kingdom led the way with legislative action, such as the UK Factory Act of 1802. Whereas, as Mark Aldrich (1997) notes, the US contribution was uniquely shaped by its cultural emphasis on rugged individualism. This ideology, which celebrated self-reliance and distrust of government, gave rise to the iconic campaign placing signs on mirrors that said "You are responsible for safety" This approach led to a period where American workplaces were often more dangerous than their European counterparts. This is still true today.

However, the export of this ideology accelerated in the 1970s and 1980s when American individualism was commoditized. Consulting firms packaged behavioral psychology into "Behavior-Based Safety" programs and sold them worldwide. This created a clash of cultures, as defined by Hofstede's (2001) dimensions of culture: multinational corporations began enforcing these individual-focused observation programs on their subsidiaries in Japan, Latin America, and Scandinavia. Collectivist cultures, which naturally gravitated toward shared responsibility, were forced to adopt an alien framework that viewed risk through the lens of individual choice (Dekker, 2019; Hopkins, 2006).

The spread and intensification of individualism produced significant negative consequences that affect accident prevention efforts. As autonomy and self-expression become overriding goals, social ties become looser. As a result, people experienced smaller and less satisfying support networks and higher levels of loneliness, hopelessness, and mental health concerns, especially among the young and elderly (Nickerson, 2023). All these results are antithetical to occupational health and safety (OHS).

This American influence is now enforced not just by culture but by contract. The US Occupational Safety and Health Administration definitions of "recordability" have become the global currency of safety. To secure contracts with

US multinationals, vendors in Germany, India, or Australia are often required to report their total recordable incident rate. This forces organizations worldwide to manage safety based on individual injury counts rather than system health, simply to satisfy the insurance and liability models of the United States (Manuele, 2013).

Another blow to the viability of individualism is that it does not deliver on its promises. Along with an iron will, an individual is expected to have the ability to save themselves. The concepts of rugged individualism and the self-made person are meant to inspire people to strive for a higher place in society, essentially to make more money. Yet, the reality is that most wealthy people come from families with wealth (Federal Reserve, 2015).

It promises that you will become fully autonomous and no longer be exploited by bosses, a reality that is impossible for the majority of people in dominant Western cultures. Ironically, individualism has become a collective belief system that legitimizes elite power. If the collective did not embrace individualism, the few could not rise to power. One person alone cannot run a company or a country; they must have a group around them to accomplish it.

The elite group then decides who can be a member. An OHS advisor shared a shocking case where his boss, who was part of the executive circle, changed the reporting dates for safety statistics to exclude a series of accidents that happened right before the end of the reporting period. That way, the executive group could still qualify for the performance bonus. The executive's decision to hide injuries is a clear demonstration of extreme individualism because the manager put their own financial reward ahead of the collective safety of the workforce. A sense of "us versus them" was created. The OHS advisor that shared this story with me asked, "Why would anyone report incidents honestly after that?"

Where a "them and us" culture exists, social cohesion has eroded. When that is gone, there is a reduced sense of shared vulnerability and collective responsibility for safety. Accidents might be seen more as individual failings rather than systemic issues requiring collective solutions. People struggling under these conditions may be less likely to report hazards or seek help when needed.

Individualism's Hidden Relationship with Mental Health

The most urgent and often overlooked connection between accidents, individualism and fatalities lies in its impact on mental health. Recent data from Eurostat (2024) reveals that nearly **one-quarter (24.7%)** of all workplace accidents are primarily driven by physical or mental stress.

Other research indicates that excessively individualistic values are associated with higher levels of isolation, suicidal tendencies, and poorer psychological health. By framing psychological well-being as a matter of personal responsibility, the system silences the very signals needed to prevent catastrophe. Research confirms that excessively individualistic values are consistently associated with higher levels of social isolation, suicidal tendencies, and overall poorer psychological health (Humphrey & Bliuc, 2022; Whitely, 2017).

Workplace stress, depression, anxiety, long working hours, job insecurity, and workplace bullying are not merely discomforts; they are recognized as significant psychosocial risk factors. In Europe, psychosocial risks are linked to approximately 10,000 deaths annually, including suicides caused by work-related depression and coronary heart disease. These mental health issues, often exacerbated by the pressures of individualistic cultures (intense competition, the burden of self-reliance, fear of failure), directly result in a substantial number of workplace fatalities, including suicides (Conway et al., 2022; Vigna et al., 2025).

While the direct classification of work-related suicides varies by country, psychosocial risks like work-related depression and stress—which are exacerbated by individualistic cultures—are now recognized by bodies like the WHO (2022) and in Europe as contributing to thousands of deaths annually, making them a vital, if often overlooked, category of workplace fatality. Beyond the immediate workplace, individualism's impact extends to broader public health crises, offering further insight into its potential for fatal outcomes in ways not typically examined in occupational safety (Jiang et al., 2022).

Recognizing the Importance of Emotions and Feelings

One of the detrimental effects of individualism is that it has culturally evolved to disregard emotions and feelings as data supported by behavioral psychology. People are urged to leave their emotions at the door when entering the workplace. Not only is this impossible, but OHS advisors know the reality of the workplace is filled with emotions, feelings, relationships, and both distrust and trust.

RCL Belief #8: *Relationships influence emotions, feelings and beliefs, which influence decisions*, is backed by research yet this belief has yet to be adopted mainstream.

Barsade (2014) published studies that quantify the influence of emotions in the workplace and on business outcomes. She found that when employees were able to express caring, compassion, and affection, it was correlated with lower burnout and stress levels. Her studies took place in nursing homes where they measured tenderness, compassion, affection, and caring. They did not ask participants if they felt or expressed those emotions themselves; the researchers asked to what degree people saw their colleagues expressing them.

They also measured employee withdrawal and their levels of emotional exhaustion by studying their rates of absenteeism. They found that units with higher levels of compassionate love had lower levels of absenteeism and employee burnout. It also showed up in higher levels of employee engagement through employee satisfaction surveys. Barsade's website notes,

> The view that dominated our field for 20 years was that anytime you engage in emotional labor — meaning you're changing or regulating your emotions for a wage —that's going to lead to burnout. What we're suggesting is that it's more complicated than that. It may well be that even if you don't start out feeling the

culture of love — even if you're just enacting it — it can lead to these positive outcomes.

The evidence is mounting that emotions such as feeling included are an active influence in organizational success and failure and that psychological injuries are as real as physical ones (Eisenberger et al., 2003). Thus, reinforcing that the way we treat people matters. To further understand these intangible forces, there are at least three sides to examine. There is the manager who may have the power to make changes in the workplace. There is the safety practitioner who may be frustrated because they do not have the authority to make changes happen. Then there is the person doing the work who might have a completely different perspective on what needs to change.

Emotions are constantly shifting, as are the perceptions of the people interacting. Can understanding the dynamic between individualism and collectivism help to explain some of these conflicts? And more importantly, can understanding their origin contribute to positive organizational change?

Consider that the safety problems and solutions are not individual but are shaped by one's role and position within the organization. The difficulties people have working together are often sparked by perceptions that others' behavior is irrational, stemming from political motivation or simply a conflict of values (Long, 2019). Focusing on the relationships and interactions between people—rather than individual traits or actions—is a collectivist approach, which might also be called a social systems approach.

This approach is esssential for leaders responsible for identifying, explaining, and addressing organizational challenges. The ability to understand the emotions of others (social perception) is highlighted as one of the strongest predictors of senior managers' effectiveness and their influence on safety (Fruhen et al., 2014).

The profound influence of social systems on individual and collective behavior is best understood through qualitative, in-depth analysis. Purely quantitative approaches cannot capture the intricate web of relationships, norms, and power dynamics that shape human action within a given social context. Unfortunately, OHS is dominated by surveys that produce numbers but very little of the context that would guide deeper interventions. The irony, as will be discussed later, is that the deeper intervention is often much simpler than the superficial one which usually involves a number of corrective actions.

By exploring narratives, interpretations, and lived experiences, qualitative analysis uncovers the subtle yet pervasive ways in which social structures, cultural values, and institutional arrangements guide, constrain, and even determine individual choices and societal outcomes.

The Illusion of Autonomy

The concept of "self-leadership" is often presented as a positive manifestation of individualism in safety. Defined as a cultural concept characterized by high individualism and low power distance, self-leadership emphasizes personal responsibility, innovation, and effective communication. Proponents suggest it can lead to improved safety climates and outcomes by empowering individuals to take initiative (Luria et al., 2016).

While meant to be empowering, self-leadership it can devolve into making safety primarily an individual's responsibility. This replaces "You are responsible" with the seemingly empowering but equally individualistic instruction to "Lead your own safety."

Self-leadership is exclusive. It is explicitly designed to flourish in high individualistic and low power distance cultures. This immediately excludes the majority of the world's countries, focusing primarily on the most wealthy, Western nations. This narrow applicability highlights its inadequacy as a universal safety model and suggests a cultural bias that overlooks the collaborative safety needs of diverse workforces.

While individual agency is highly prized in individualistic cultures, it does not necessarily correlate to fewer accidents, as has been shown in numerous surveys where 100% of employees assert their personal responsibility for safety while the accident statistics tell a different story.

Furthermore, self-leadership, like empowerment, when poorly defined, can lead to a loss of trust. This happens when an employee makes a decision that management does not approve of and retracts their authority.

A Preference for Independence

Even in individualistic societies, individual actions have underlying social causes. For example, career choices, personal values, and even the expression of individualism itself can be shaped by social expectations, cultural norms, or economic opportunities provided by a particular group. Durkheim poignantly points out that even the most individualistic action we can take is grounded in social needs. He was referring to suicide (Jones, 1986).

Our preference for thinking of ourselves as independent individuals, driven by our own internal motivations and choices, is deeply ingrained in Western cultures. Yet, this preference often clashes with the reality of our social nature and the profound influence of social forces on our behavior. This is precisely why a relationship-centered approach to safety can meet with resistance; it requires acknowledging that we are not as autonomous as we might like to believe.

Research in social psychology consistently demonstrates that very few individuals defy authority or group norms. The original studies by Asch (1955), Milgram (1963), and Zimbardo (1970) show that individuals are highly susceptible to social pressures, often conforming to clearly wrong group decisions and obeying authority figures even when it goes against their personal values. More recent studies confirm these findings (Burger, 2009; Franzen and Mader, 2023).

Milgram's experiment shockingly revealed participants' willingness to inflict pain on others simply under instruction from an authority figure. Zimbardo's Stanford Prison Experiment further illustrated how artificial social conditions could evoke cruel behaviors not previously exhibited by participants. These findings challenged established beliefs about free will and self-determination, opening the way to question the expectation that individual workers should adopt ways of working that their peers do not support. This research raised a lot of questions about what society believed about free will and self-determination up until then. It questioned the belief that individuals have complete control over their actions and can resist any external pressure if they strengthen their character.

Many struggle to accept the findings of these experiments. It is distressing to contemplate that one might violate personal values and succumb to authority. It is more comforting to believe in oneself among the small minority immune to such pressure. However, transformational change begins within the leader. That means taking a hard look at what we believe to be true and being willing to let go of mistaken beliefs.

For the OHS advisor, this research helps to explain why people do things like stick their hands into moving machinery or fail to wear required protective equipment. It can only be understood and addressed by becoming aware of the psychological influences on how people make decisions. Deviations from procedure can be due to peer-group pressure or the expectations that peers and management have set for what is acceptable at work. Once this is understood, it is easier to understand why OHS must shift its focus from attempting to control individual will to understanding and influencing the social context that shapes risk decisions. It also explains why re-training or updating procedures is not enough to prevent accidents. It is a group dynamic, so any attempt to change this behavior must be done at the group level.

Change the focus from individual will to social context

Dr. Long, a longtime researcher and advocate of the primacy of social psychology in risk assessment, says that the notion of individual will—the ability to override cultural heritage, history, social influence, or emotions—stems from a Cartesian worldview. He asserts that contemporary research in neuroscience, cognition, and social psychology unequivocally demonstrates that the concept of individual will is unfounded. This understanding is essential when addressing risk within organizations. While we may believe we can consciously control our attention, this is not accurate. Therefore, Dr. Long suggests that OHS should shift its focus from individual will to the social context surrounding the worker, including their worldview and the collective unconscious, as these are the true sources that influence risk decisions (Long, 2019; Pupulidy and Vesel, 2023).

While the goal is not to reject individual thought, focusing solely on individual awareness can be harmful. This is a difficult point to argue because people raised with individualistic values often feel threatened by the idea that groups can influence their thinking. Niklas Luhmann's (1982) framework helps clarify this: he differentiates between individual "thinking systems" and "social systems" (groups). The latter produce communication, while individuals generate thoughts and self-awareness.

Luhmann argued that while individuals are not fully controlled by groups, they are also not entirely separate, making them susceptible to social influence. As demonstrated by the Milgram and Asch experiments, this susceptibility varies. It is important to understand that resisting group influence does not imply moral superiority; rather, the expectation that everyone should conform to individualistic ideals hinders our understanding of accident causation. This is a core reason that excessive individualism can be detrimental to safety.

Adding to the research on the importance of social connection, neurological research through brain scans has shown that social rejection and exclusion trigger the same pain pathways in the brain as physical pain. In an experiment, participants played a game on a screen in an MRI scanner. At first, all three players received the ball, but after a while, only two began to play with each other, leaving out the third party who was the experimental subject. The scans showed that the part of the brain lighting up was the same part that lit up when people felt rejection or pain (Eisenberger et al., 2003). This surprising biological connection underscores that the threat of social rejection can be as powerful, if not more so, as the threat of physical harm. This finding fundamentally impacts safety, emphasizing that creating an environment where people feel included and that they belong is imperative to their well-being, while preaching that they alone are responsible for safety may reduce people's capabilities.

Having established the deep-seated social nature of human beings, we will now explore how safety management has evolved and where it has failed to account for these human social and psychological dynamics.

Conclusion

We have traced the roots of the "sovereign individual" from the Era of Enlightenment to the modern factory floor. The verdict is clear: autonomy from social pressure is rare. Research from Asch, Milgram, and modern neuroscience proves that we are biologically wired for connection, not isolation. This is a significant insight for safety, as it proves that a system that manages safety at the individual level is inherently flawed.

By designing safety systems for individuals who rarely exist, we fail the people who do. We cannot manage safety at the individual level because safety does not live there. It lives in the space between us. The task now is to build a system that reflects that reality.

The chapters that follow will build on this foundation by exploring the practical tools and frameworks needed to cultivate a culture of psychological safety, collective responsibility, and genuine human connection.

References

Aldrich, M. (1997). *Safety First: Technology, Labor, and Business in the Building of American Work Safety, 1870–1939*. Johns Hopkins University Press.

Asch, S. (1955). *Social Psychology*. HarperCollins.

Barsade, S. (2014). *Why Fostering a Culture of Compassionate Love in the Workplace Matters*. Knowledge at Wharton staff. https://knowledge.wharton.upenn.edu/article/fostering-culture-compassion-workplace-matters/
Bauman, Z. (2000). *Liquid Modernity*. Polity Press.
Burger, J. R. (2009). Replicating milgram: Would people still obey today? *American Psychologist*, 64(1): 1–11.
Bush, C. (2021). *Preventing industrial accidents: reappraising H. W. Heinrich*. Taylor Francis.
Conway,P. M, Erlangsen, A., Grynderup, M. B., Clausen, T., Rugulies, R., Bjorner, J. B., Burr, H., Francioli, L., Garde, A. H., Hansen, Å. M., Hanson, L. M., Kirchheiner-Rasmussen, J., Kristensen, T. S., Mikkelsen, E. G., Stenager, E., Thorsen, S. V., Villadsen, E., & Høgh, A. (2022). Workplace bullying and risk of suicide and suicide attempts. *Work Environment Health*, 48(6): 425–434. https://doi.org/10.5271/sjweh.4034
Dekker, S. W. A. (2019). *Foundations of Safety Science: A Century of Understanding Accidents and Disasters*. Routledge.
Dwyer, T., & Raferty, A. E. (1991). Industrial accidents are produced by social relations of work: A sociological theory of industrial accidents. *Applied Ergonomics*, 22(3): 167–178. https://doi.org/10.1016/0003-6870(91)90156-C
Eisenberger, N.I., Lieberman, M.D.,&Williams, K.D. (2003). Does rejection hurt? An fMRI study of social exclusion. *Science*, 302(5643):290–292. https://doi.org/10.1126/science
Eurostat. (2024). *Accidents at work statistics*. https://ec.europa.eu/eurostat/statistics-explained/index.php?title=Accidents_at_work_statistics
Federal Reserve Bank of St. Louis. (2015). *An overview of the connection between education and wealth*. www.stlouisfed.org/on-the-economy/2015/july/education-correlation-wealth-income-video
Flin, R. (1996). *Sitting in the Hot Seat: Leaders and Teams for Critical Incident Management*. Wiley.
Flin, R., Mearns, K., O'Connor, P., & Bryden, R. (2000). *Safety Climate: A Review of the Literature*. Offshore Safety Division, Health and Safety Executive.
Flin, R., O'conner, P., & Crichton, M. (2008). *Safety at the Sharp End*. Ashgate.
Foucault, M., & Gorz, A. (1989). *Critique of Economic Reason*. Verso Books.
Franzen, A., & Mader, S. (2023). The power of social influence: A replication and extension of the Asch experiment. *PLoS ONE*, 18(11): e0294325. https://doi.org/10.1371/journal.pone.0294325
Fruhen, L. S., Mearns, K. J., Flin, R., & Kirwan, B. (2014). Skills, knowledge and senior managers' demonstrations of safety commitment. *Safety Science*, 69: 29–36.
Hofstede, G. (2001). *Culture's Consequences: Comparing Values, Behaviors, Institutions and Organizations across Nations* (2nd ed.). Sage Publications, Inc.
Hopkins, A. (2006). What are we to make of safe behaviour programs? *Safety Science*, 44(7): 583–597.
Humphrey, A., & Bliuc, A.-M. (2022). Western individualism and the psychological well-being of young people: A systematic review of their associations. *Youth*, 2(1): 1–11.
Jiang, S., Wei, Q., & Zhang, L. (2022). Individualism versus collectivism and the early-stage transmission of COVID-19. *Social Indicators Research*, 164(2): 791–821.
Jones, R. A. (1986). *Emile Durkheim: An Introduction to Four Major Works*. Sage Publications, Inc.
Long, R. (2019). *The Social Psychology of Risk Handbook: i-thou*. Scotoma Press.
Luhmann, N. (1982). *The Differentiation of Society*. (Trans. Stephen, H., & Charles, L.). Columbia University Press.

Luria, R., Balaban, H., Awh, E., Vogel, E. (2016). The contralateral delay activity as a neural measure of visual working memory. *Science Direct*, 62: 100–108.

Manuele, F. A. (2013). *On the Practice of Safety*. Wiley.

Milgram, S. (1963). Behavioral study of obedience. *Journal of Abnormal and Social Psychology*, 67: 371–378.

Nickerson, C. (2023). Individualistic cultures and example behavior. *Simply Psychology*. www.simplypsychology.org/what-are-individualistic-cultures.html

Peetz, D. R. (2010). *Are Individualistic Attitudes Killing Collectivism?* Griffith University.

Pupulidy, I., &Vesel, C. (2023). Human & Organization Potential. Dynamic Inquiry LLC.

Unesco (2024). Human rights, liberation and reconciliation: Nelson Mandela legacy sites. Unesco world heritage convention. https://whc.unesco.org/en/list/1676/

Vigna, E., Balkom, I.V., Bresnahan, M., Cheslack-Postava, K., Musa, G., Ryan, M., Skokauskas, N., Hoven, C., &Carli, V. (2025). Unravelling the threads: Understanding the interplay of cultural values, female workforce engagement, human development index and suicide rates. *Archives of Women's Mental Health*, 28(2): 287–296. https://doi.org/10.1007/s00737-024-01502-9

Whitely, R. (2017). Is an increase in individualism damaging our mental health? *Psychology Today*. www.psychologytoday.com/us/blog/talking-about-men/201707/is-increase-in-individualism-damaging-our-mental-health

WHO. (2022). *Mental health at work*. www.ilo.org/sites/default/files/wcmsp5/groups/public/@ed_protect/@protrav/@safework/documents/publication/wcms_856976.pdf

Zakaras, A. (2022). *The Roots of American Individualism: Political Myth in the Age of Jackson*. Princeton University Press.

Zimbardo, P. (1970). *The Lucifer Effect: Understanding How Good People Turn Evil*. Random House.

4 The Evolving Nature of Safety Management

Introduction

We began by unmasking the profound consequences of the hyper individualism that is affecting our world and move on to introducing the social skills and principles that will further safety's movement towards a relationship-centered approach. The shift will not easily take place. The occupational health and safety (OHS) profession has come to perceive safety as an individual pursuit. That very perception is a barrier to the type of collective approach that could save lives.

The Current Crisis

We are in a state of crisis not just with workplace serious injuries and fatalities (SIFs) but in every aspect of our lives that determine mental, physical, and financial wellness. Two of our most respected social commentators, David Brooks and the late Mark Fisher, are in alignment about linking the crisis to individualism. Brooks identifies a "hyper-individualistic culture" as a direct contributor to a crisis of social connection, citing alarming statistics such as a 45% rise in suicide rates linked to isolation. Fisher critiqued what he termed "mandatory individualism," where systemic issues, including what he called a "pandemic of mental anguish," are wrongly framed as private problems to be solved by the individual on their own (Reynolds, 2017).

Their shared conclusion is that forcing people to face complex challenges without support degrades well-being and prevents the collective action needed to solve systemic problems. This societal trend directly impacts the health of our workforce.

The *Social Animal* (Brooks, 2012) argues that humans are inherently shaped by the unconscious mind, which is deeply emotional and driven by a primal biological need for connection. The book synthesizes research from neuroscience and psychology to illustrate that success and happiness are emergent properties of our relationships. He concludes that our brains are wired to connect with the minds of others rather than function in isolation. This goes back to the issue of just how independent we are from the influence of others. The individuals who influence

DOI: 10.4324/9781003668770-4

us—supervisors, friends, colleagues, and family —contribute to the thoughts we often perceive as solely our own.

These critiques are not seeking to eliminate individualism but an attempt to light a path toward inclusion that also allows creative freedom. Mark Fisher wrote in 2013.

> We need to learn, or re-learn, how to build comradeship and solidarity instead of doing capital's (the ruling class) work for it by condemning and abusing each other. This doesn't mean, of course, that we must always agree – on the contrary, we must create conditions where disagreement can take place without fear of exclusion and excommunication.
>
> [Original published in 2013]

The critique of hyper-individualism and the associated destruction of social connection does set the stage for a deeper exploration into how the dominance of individualism shapes safety outcomes in the workplace. Respected researchers like Gene Rochlin found the connection by studying the influence of social practices and collective knowledge on operational safety.

Influences Changing the Face of Risk Management

Gene Rochlin (1999) identified the importance of "local knowledge" and the necessity of social practices for ensuring safety. He challenged the "human error" paradigm, arguing against the customary focus on individual fallibility as the primary cause of accidents. He demonstrated how technology and social practices are intertwined, and how changes in one can disrupt the other, creating new safety challenges.

Rochlin's work underscored that safe operations in high-hazard industries depend on established social practices and practical knowledge among workers and that disrupting these practices, even with well-intentioned technological interventions, can inadvertently elevate risk. Rochlin emphasized the invaluable nature of local knowledge and expertise developed through experience and interaction—a tacit understanding vital for safe operations.

This chapter marks a pivotal juncture in the evolution of safety management as a social psychology discipline (Long, 2019). However, there is a strong resistance from a segment of the profession that wants to maintain compliance and engineering as the focus of safety management. The argument is that the existing body of knowledge and controls is sufficient if only properly implemented, and that socio psychological issues are a distraction. To them the real issues are managers and workers who disregard safety protocols for their own benefit.

Social psychology pursues questions that have baffled the safety industry for decades: Why do incidents like reaching into operating machinery still occur despite safety measures? It is not because employees, managers, supervisors, or safety advisors are negligent or uncaring We need a new line of inquiry and we need to implement tools that are mostly ignored such as (Figure 4.1) Non-Technical Skills.

As safety evolves from a technical pursuit to a recognized social science, it prompts a deeper inquiry into the human competencies that influence safety outcomes—the intricate dynamics of social interaction, cognition, and communication. Within this context, three critical but often conflated concepts emerge as central: cognitive, relational, and personal resource management skills (Figure 4.1).

These skills are the essential social competencies required for individuals to navigate and positively influence both social and technical systems. It is relational skills—like communication, teamwork, and leadership—that **expand the workforce's cognitive and social capacity to identify and report risk.**

NTS are not about technical knowledge or manual dexterity but rather about understanding how individuals and teams perform under pressure, anticipate risks, and respond effectively to unexpected events (Arpansa, 2025). The research on the direct impact of well-developed NTS on preventing accidents and optimizing performance underscores their importance in safety management and operational success (Army University Press, 2022).

NTS began with the study of airplane crashes and now have a long record of success in reducing them. They were developed from what was learned from black box recordings after a plane crashed. Investigators noted consistently that the captain was either not listening or failed to elicit information from the rest of the crew during a crisis. More critically, sometimes a person in the cockpit raised a concern or observation that might have prevented the crash, but the captain did not act on it. Upon further investigation, it emerged that the lower status of the co-pilot sometimes made them afraid to speak out, and contributed to why the captain did not act on it. The training developed to address this socio-psychological problem became known as Crew Resource Management (CRM).

NTS as the Tool for Social Construction

If "Social Construction" is the house we live in (our shared reality of risk), and "Sensemaking" is the act of building that house, then NTS are the tools used to build it. These skills are now taught in all high risk industries as they are specifically tailored for high-stakes operational tasks where human error can have severe consequences.

By honing these skills, professionals significantly enhance the application of their technical knowledge. However, NTS do not function in a vacuum. NTS provide the mechanism for exchange, but trust provides the medium. Without the relationship, the skill is useless.

The Executive Mandate: Modeling NTS

An executive's responsibility is not to become a master trainer in NTS but to champion and model these behaviors (King. 2023). Research confirms that an executive's

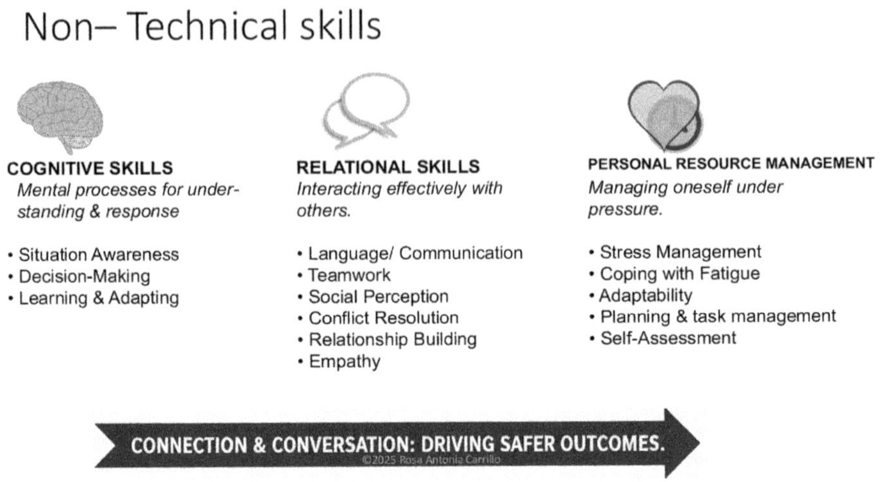

Figure 4.1 Non-Technical Skills.

commitment to safety is most visible not through their technical directives but through their demonstration of NTS—specifically their social and problem-solving skills (Fruhen et al., 2014). When a leader models good NTS, they validate it as a critical competency for the entire organization.

Summing Up NTS and CRM

We have been emphasizing how to approach the social and psychological factors in risk management. Yet the biggest challenge is the belief of individual invulnerability. Pilots from all regions of the world have been found to hold unrealistic attitudes about the effects of stressors on their performance—the majority feel, for example, that a truly professional pilot can leave personal problems behind while flying and that their decision-making ability is the same in emergencies and normal operations. In summary:

> CRM is not and never will be the mechanism to eliminate error and assure safety in a high risk endeavor such as aviation. Error is an inevitable result of the natural limitations of human performance and the function of complex systems. CRM is one of an array of tools that organizations can use to manage error. The safety of operations is influenced by professional, organizational, and national cultures and safety requires focusing each of these toward an organizational safety culture that deals with errors non-punitively and proactively.
>
> (Merritt & Helmreich, 1999:9)

Language Shapes Reality

Chapter 7 provides a deep dive into why and how language impacts risk perception. Here it is introduced as part of the evolution of our understanding about human nature and how to work with it rather than try to change it.

Language is not a soft skill; it is a structural element of NTS. The specific words chosen to deliver a message shape the reality of the listener. Therefore, the language used during operations deserves significant attention from leaders.

Language is a vital tool for safety as it evolves beyond purely technical or individualistic approaches towards the social sciences. In addition, and particularly relevant to safety, the global shift to remote and virtual work has transformed the nature of interpersonal skills and their application. The absence of physical proximity and conventional non-verbal cues has elevated certain NTS from "nice to have" to essential for survival. AI and automation are further accentuating the need for these uniquely human competencies, making them key differentiators for professional success in a rapidly changing world.

Case Study: Socio-Technical Systems

It was not the intention of these two social psychologists to create a new framework to understand the interface between people and technology, but that is what happeded and it came to be called the socio-technical systems organizational framework.

In 1963, Emery and Trist were brought in to solve a production problem in the British coal mines. The mines had just been reorganized and new equipment installed to help increase production, but things were not working out. Absenteeism rose, production fell, and accidents happened more often. As is often the case, the technology had been installed without a prior investigation into its consequences and benefits for the workers.

The miners explained that, traditionally, they had shared all tasks, preventing anyone from being stuck with the same job all day, and they were empowered to organize their work. All this self-management was lost with the new structure, resulting in significantly lower morale. This decline was further evidenced by the breakdown of their social bonds: workers were no longer socializing or caring for one another outside of work. Previously, if a coworker was sick or injured, the others would cover for them and even bring them food (Trist et al., 1963).

After observing the miners and speaking with them, Emery and Trist delivered their findings. The key finding was that the new, assembly-line-inspired organizational structures were mismatched with the realities of underground work. Absenteeism rose and production fell because the "social system" (relationships) was the actual operating system, not the machinery. The traditional, cohesive workgroups, with their end-to-end responsibility and ability to make on-the-spot decisions, were far better equipped to handle the uncertainty and complexity of the mining environment.

This research established the fundamental principle that an organization operates best when there is a good match for the people and the technology that they use. This became known as the STS theory.

Thus history shows, from the early mines in the UK to the Deepwater Horizon disaster, that safety failures occur when the 'technical' is prioritized over the 'social.' A successful socio-technical system requires that we treat people not as a risk to be managed, but as the essential 'match' that enables technical systems to function safely and reliably (Roberts & Bea, 2001).

This was the beginning of looking at a successful operation as a social construct. As Rochlin said decades later, safe operations in high-hazard industries depend on established social practices and practical knowledge among workers. Disrupting these practices, with well-intentioned technological interventions, can inadvertently elevate risk.

The Enduring Challenge: Individualism and Collective Imbalance

The enduring challenge is that organizations often fail to consider the relationship between the social (people) and the technical (technology). This failure is frequently manifested through the dominant ideology of individualism. In this outdated mindset, workers are diminished and often viewed as predictable entities—physical assets that are not valued as highly as structural assets. This perspective was captured by a high-level manager who admitted he knew everything about his physical assets and very little about the people who worked for him.

The consequence of this imbalance is the persistent stagnation of SIF rates. The shift to STS today is resurfacing precisely because of the need to move toward a collective approach that acknowledges that the workplace is primarily social. So there must be the same attention placed on people as there is on physical assets.

Can Safety Evolve into Becoming People Centric?

The concern today is focused on fatalities because the number of accidents is going down while fatalities rise. Safety practitioners are often divided on the best approach to lower SIFs. Some believe everything needed to prevent incidents is already provided by regulations and current safety management systems. They say the problem is that companies do not enforce and comply with them (Bush, et.al., 2020)

Another group points to the unpredictable nature of risk and advocates for a new paradigm mostly centered on enabling employees to employ their skills and knowledge to prevent incidents. Introducing that premise creates a polarity—a conflict that is unresolvable and can only be managed. This one is "human-centric vs compliance." One of the most frequent ones in industry is investing in "people vs physical assets." Such issues cannot be resolved once and for all, so they never get resolved (Carrillo, 2005).

One of the polarities triggered by the socio-psychological framework is "compliance vs adaptation." Can risk be managed more effectively by exerting controls or by providing the education to allow for flexible adaptation? This is a polarity that is not likely to resolve itself in favor of one view over the other because they are both true.

Barry Johnson (1994) called it "polarity management." He distinguishes a polarity from a complicated problem, which can be solved. Also needed is a high tolerance for ambiguity because the conflict never goes away. The reason is that emphasizing one aspect of the polarity makes the other side worse. For example, think about how many times companies have centralized safety resources to gain better control, only to get the feedback that they need to decentralize them again because the business units do not have enough OHS support. When the shortcomings of decentralization appear, the department is again centralized, and so on.

This topic is very relevant currently because there is a lot of conflict on social media sites between OHS practitioners that support emphasizing compliance and those who propose that a human centric approach is more effective. The debate can become quite acrimonious. This conflict not only keeps the OHS occupation from having a unified identity but it is also one of the most common frustrations that OHS practitioners say causes personal burnout and dissatisfaction. This will be further discussed in Chapter 13, the advisor as a risk navigator.

Figure 4.3 Figure 4.2

The polarity management model (Figure 4.2) can be used to visually compare the two sides of the conflict. It communicates that both sides have positive and negative outcomes. It gathers the contributions each side feels that they make and what each side fears will happen if the other dominates. This is where the conflict resides. Once the fears are spoken, there is an opportunity to create a shared philosophy for how to improve safety and reduce fatalities. You might say that this is a sense-making process.

Fears the Human-Centric Side Has about Traditionalists

This group is concerned with the human and systemic aspects of safety, and they fear that rigid, top-down control will hinder an organization's ability to learn and evolve. The source of conflict is their focus on these fears.

- **Failure to Learn:** They fear that a sole focus on compliance will lead organizations to repeatedly miss underlying social issues. They worry that a company will blame a "bad apple" or a "procedural violation" for an incident, rather than investigating the complex human and organizational factors that contributed to it.
- **Lack of Flexibility:** They worry that strict rules and procedures will force workers to choose between following the rules and getting the job done safely. They know that real work is messy and unpredictable, and a rigid system can't account for the on-the-ground reality.
- **Unexpected Consequences:** People will do workarounds and cut corners to fit things in (Flin et al., 2008).

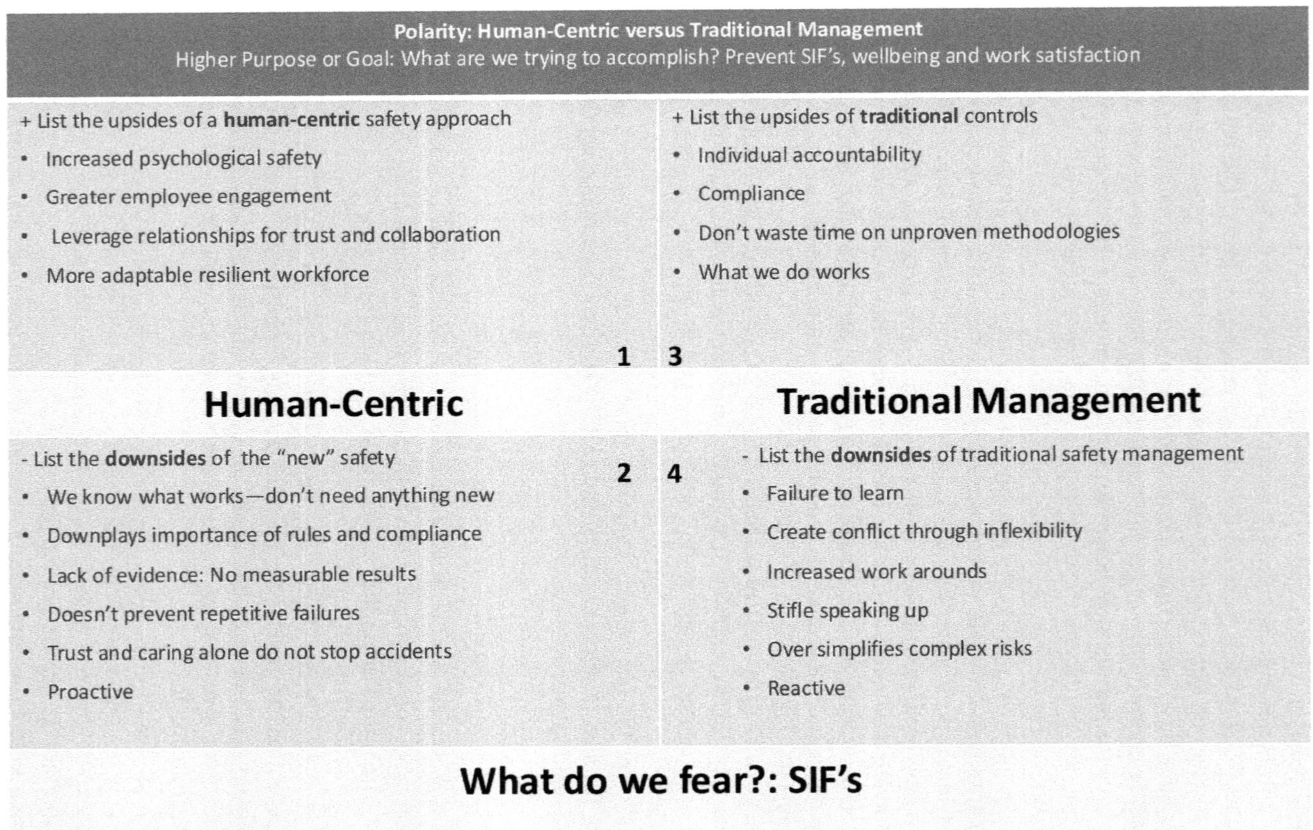

Figure 4.2 Balancing Human-Centric with Traditional Safety Management.

- **Stifling of Dialogue:** They fear that a compliance-heavy culture will discourage workers from speaking up about close calls or offering suggestions for improvement. If the emphasis is on punishing non-compliance, employees will hide problems rather than share them.

Fears the Traditionalists Fear about the Human-Centric Proponents

- **New accidents theories are not needed.** We know what works. The new safety view is distracting the safety discipline from implementing proper existing safety controls.
- **The new safety view downplays the importance of rules and compliance**, subverting years of work dedicated to making the workplace safe. Relationships and Trust alone will not stop a fatal accident. You need basic risk management and controls.
- **The new safety has failed to demonstrate measurable results.** We suspect they are simply trying to generate more consulting business.
- **After every disaster they find the same deficits.** We develop methods to address those deficits, but they are not implemented. The new safety has not shown any viable means of preventing such disasters.

The fears are acknowledged. Representatives of each perspective agree upon how to identify when one approach is showing signs of creating a problem. By actively discussing and agreeing on the warning signs that the organization has gone too far to one side of the polarity, the organization can create a feedback loop that helps them self-correct. This allows them to maintain a healthy tension between the two approaches, rather than swinging wildly from one extreme to the other. This process builds trust and shows that both sets of values are respected and necessary for a holistic safety culture.

Structuring consensus: Detailed instructions for how to facilitate a polarity mapping exercise to reach an agreement is provided in Appendix E. The purpose is to address these deep-seated conflicts and manage the polarity. Organizations need to move beyond a simple compromise and create a new framework that integrates both perspectives. This requires a shared vision and a collaborative process.

Conclusion

The tension between compliance/control and adaptation/relationships is unsolvable and must be continuously managed. The most effective approach to safety management is to manage it as a polarity (a conflict or dilemma that cannot be resolved). Overcoming stagnation requires a "Both/And" strategy that integrates compliance and adaptation.

We are stuck in a false war between "Compliance" and "adaptation," between "efficiency and thoroughness." The goal is not to choose a winner but to manage the tension. Safety lives in the messiness of the "And." It requires rules *and* relationships, and structure *and* sensemaking. The leader's job is not to pick a side but to know when to shift the weight.

Executive Takeaways

- Utilize the polarity management concept to introduce the idea that some thorny issues cannot be resolved. They can only be managed. Use. the concepts of the compliance vs adaptation polarity to lead a discussion with your leadership team. Map out the tension between your organization's need for strict rules and its need for flexible, human-centric solutions. The goal is to develop a "Both/And" strategy not an "Either/Or" one. The procedural steps for mapping that conflict can be delegated to an OHS practitioner or experienced facilitator using the guide in Appendix.
- Evaluate your leadership team's skills against the "new set of skills" mentioned in this chapter (non-technical, soft, and socially literate). Invest in development programs that focus on social perception, empathy, and communication to build the foundation of trust that enables compliance to work.
- Begin consciously shifting the corporate vocabulary from one of enforcement and individual accountability to one that includes relationships, trust, shared understanding, and collective responsibility. This models the behavior you want to see and reinforces the values of a holistic safety culture.

References

Army University Press. (2022). *Nontechnical Skills.* Army University Press. Accessed September 7, 2025. www.armyupress.army.mil/Journals/Journal-of-Military-Learning/Journal-of-Military-Learning-Archives/April-2022/Nontechnical-Skills/

Arpansa. (2025). *Non-Technical Skills.* Australian Government. www.arpansa.gov.au/regulation-and-licensing/safety-security-transport/holistic-safety/non-technical-skills

Brooks, D. (2012). *The Social Animal: The Hidden Sources of Love, Character and Achievement.* Random House.

Carrillo, R. A. (2005).Safety leadership: Managing the paradox. *Professional Safety Journal*, 7: 31–34.

Fisher, M. (2013). Exiting the Vampire Castle. *Open Democracy.* www.opendemocracy.net/en/opendemocracyuk/exiting-vampire-castle/

Flin, R., O'Conner, P., & Crichton, M. (2008). *Safety at the Sharp End.* Ashgate.

Fruhen, L., Mearns, K., Flin, R., & Kirwan, B. (2014). Skills, knowledge and senior managers' demonstrations of safety commitment. *Safety Science*, 69: 29–36.

Helmreich, R. L., Merritt, A. C., & Wilhelm, J. A. (1999). The evolution of crew resource management training in commercial aviation. *International Journal of Aviation Psychology*, 9(1): 19–32. https://doi.org/10.1207/s15327108ijap0901_2

Johnson, B. (1994). *Polarity Management: Identifying and Managing Unsolvable Problems.* HRD.

King, M. (2023). *Technical Skills Won't Be Enough for the Workplace of the Future.* Fast Company. www.fastcompany.com/90964567/why-technical-skills-wont-be-enough-for-workers-thrive-in-the-economy-of-the-future?>

Long, R. (2019). *The Social Psychology of Risk.* Scotoma Press.

Reynolds, S. (2017). Mark Fisher's K-punk blogs were required reading for a generation. *The Guardian*. www.theguardian.com/commentisfree/2017/jan/18/mark-fisher-k-punk-blogs-did-48-politics

Roberts, K., & Bea, R. (2001). Must accidents happen? Lessons from high-reliability organizations. *Academy of Management Perspectives*. 15(3): 70–78.

Rochlin, G. (1999). Safe operation as a social construct. *Ergonomics*. 42(11): 1549–1560.

Trist, E. L., Higgin, G. W., Murray, H., & Pollock, A. B. (1963). *Organizational Choice: Capabilities of Groups at the Coal Face Under Changing Technologies: The Loss, Rediscovery, and Transformation of a Work Tradition*. Tavistock Publications.

5 Safety as a Social Outcome

Introduction

The purpose of this chapter is to transition from the theory of relational safety to the application of that theory using a warehouse case study to demonstrate their vital importance. It then provides the social competencies for leaders to build relationship-centered organizations.

The chapter argues that accidents are not random events caused by "bad apples" but are outcomes produced by the ways people relate to and with each other. It also provides the socio-technical iceberg as a forensic and design tool for a Socio Technical organization.

Saying that safety is an outcome of the way people treat each other does not mean that the individual has no role in safety since they are often the one in the field catching and fixing things that come up unexpectedly (Weick, 1988; 2024). This point is further explored in the sensemaking and social construction section in Chapter 6. For now suffice it to say that most of us as individuals are never quite as independent as we would like to see ourselves. That is simply not the way humans are wired to survive.

While Chapter 4 defines NTS, this chapter exemplifies how they can fail by referencing the Fukushima disaster. It argues that a culture that suppresses open discussion renders even the best NTS training in effective.

The Warehouse Case

The following story from a warehouse safety supervisor shares his awakening from the myth of individual responsibility:

> The way we did work was that everything had to be as efficient as possible. In the warehouse, that means if you want to reduce the cost per case going out the door to increase profitability, You're going to do it at the expense of employee ergonomics (Srivastava & Mishra, 2025).
>
> New things started popping up like dislocated shoulders because you have a 5-foot gentleman trying to put a product on the top shelf. And you are filling

DOI: 10.4324/9781003668770-5

delivery trucks from floor to ceiling, so they stack pallets to reach higher. So now you have people falling off the pallets, onto concrete or asphalt.

Then when you bring these things up, use your safety science to identify the hazard, you're told to go out and make a good speech. Just tell the guys to do better and make sure they stretch before starting work. I'm no moron; I know that's not the answer.

For years I had a poster on my wall that pretty much summed up management's attitude. It said, "If you're going to be dumb, you need to be tough." So if you had an accident, you were dumb, so now you had to be tough because you wouldn't be going to the doctor.

A new boss came on and I told her I'm not sure I want to continue in safety. I just don't feel like I'm able to help people. She said, read this material. We're going to be changing things around here. It was a paper about Just Culture and not blaming employees. Everything changed, and I stayed on" (Carrillo, 2021 Conversation).

This raw account highlights several key themes we explore in this book. First, the story demonstrates how individualism presents itself in safety. The safety supervisor describes a system where individuals are held accountable for preventing their injuries without the power to change things. To make things worse, employees are threatened and told that it is their fault if they have an accident. "So if you had an accident, you were dumb, so now you had to be tough because you wouldn't be going to the doctor."

Secondly, it demonstrates how organizational priorities, such as efficiency and profitability, can create goal conflicts that have to be resolved by frontline workers making the system more vulnerable. Third, it exposes the limitations of relying on simplistic solutions like increased focus on hand injuries or generic advice like stretching, despite the supervisor's clear explanation of the ergonomic hazards. Finally, the safety supervisor's disillusionment underscores the frustration of those who recognize the inadequacy of traditional, blaming approaches and feel unable to create meaningful change.

STS Iceberg Analysis

The following analysis of the warehouse case uses the socio-technical systems (STS) Iceberg (Figure 5.1) to demonstrate its use by recording and picturing what took place.

Analysis of the Warehouse Case As Told by Supervisor

Beginning with the mental models management's mental model about risk management was "If you are going to be dumb, you need to be tough." The individualistic approach led to the "fix," or structure that they set up to prevent further injuries—stretching. The inadequate manpower and equipment was ignored. They skip over the relational iceberg layer demonstrated by the fact that the pattern and

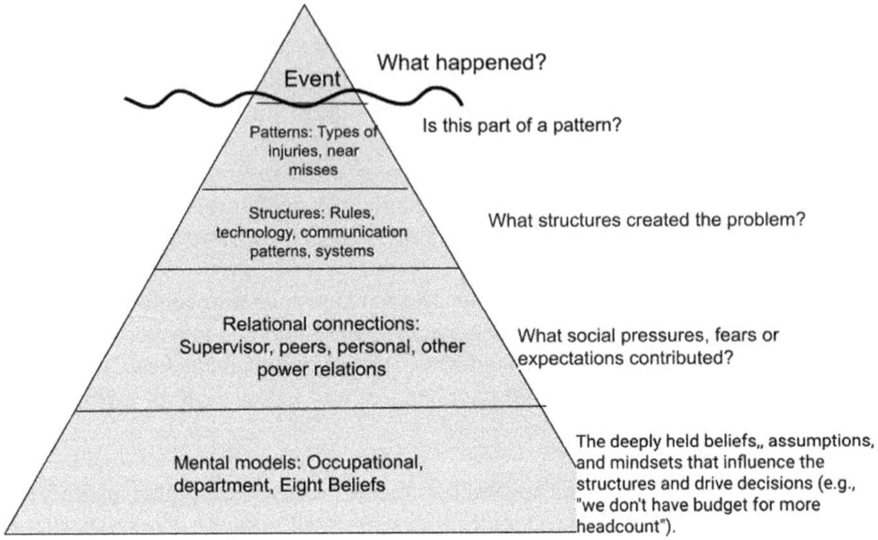

Figure 5.1 Socio-technical Iceberg.

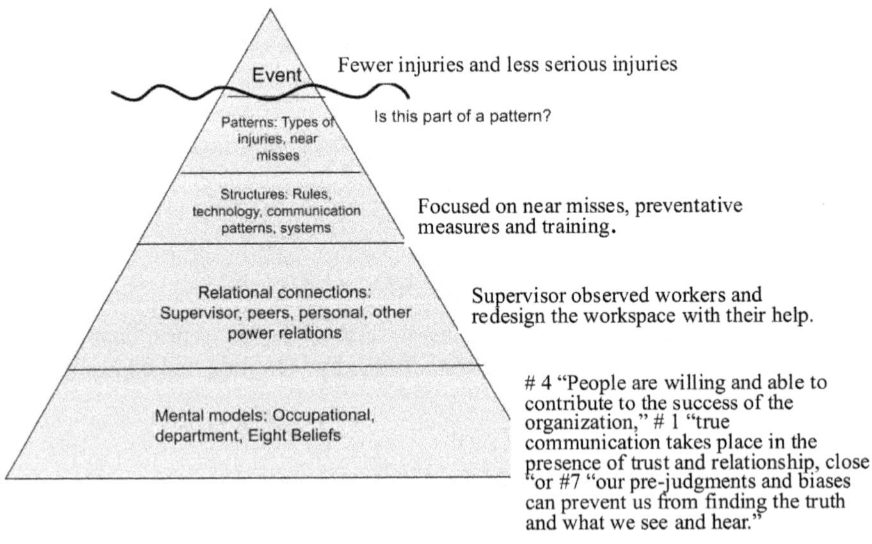

Figure 5.2 RCS Iceberg Analysis of Warehouse Case.

the cause of injuries were diagnosed by the supervisor, but he was ignored. The event or result at the top was that the injuries continued.

In contrast to the limitations of individualism, the next section presents the strength of a relational philosophy and approach that dramatically improves organizational performance while creating an environment where people thrive.

Eight Beliefs of Relationship-Centered Leadership

Over the years I have observed leaders successfully raise safety performance by treating people with respect and appreciation. Based on their behavior and subsequent conversations, I gleaned a list of eight beliefs they hold about human nature and how they assign intentions to people's behavior. The eight beliefs shown in Table 5.1 are interdependent in the sense that they all work together to build the relationships within the organization or team to encourage both collaboration and the willingness to risk bringing up interpersonal conflicts or ideas outside the norm. They also guide how the leader communicates, uncovers reality, and makes decisions in a way that maintains relationship.

When you compare management responses to the event, you can identify how the beliefs were violated and why they led to the safety supervisor wanting to quit. This example also illustrates how the intangible emotions and feelings of the workplace manifest depending on how employees feel treated. Are they trusted and valued, or are they considered unfeeling assets?

Beginning with the relational conditions, employees were clearly told they were on their own—a clear example of hyper-individualism. This approach produced the company's mental models for safety management. The individualistic approach is clearly stated in the "fix," or structure, that they set up to prevent further injuries. The pattern and the cause of injuries were diagnosed by the supervisor, but he was ignored. When you compare these responses to the eight beliefs, you can identify how they were violated.

Table 5.1 Eight beliefs of relationship centered safety

Eight Beliefs
1. True communication takes place in the presence of relationship and trust.
2. Inclusion precedes accountability.
3. Innovation, resilience, inclusion, and accountability are interdependent.
4. People are able and willing to contribute to the success of the enterprise.
5. People will speak up to stop an unsafe situation if it is in their interest to do so.
6. Drift is a positive quality of adaptive human behavior.
7. Our pre-judgments and biases can prevent us from finding the truth in what we see and hear.
8. Relationships influence emotions, feelings, and beliefs, which influence decisions.

Table 5.2 Organizational levels and change

Level	*Change management application*
Individual level	Focus: Each person within the organization. Topics: Individual psychology, behavior, reactions to policies, job satisfaction, and performance. Examples: Changes to a job assignment, disciplinary action, or a transfer.
Group or team level	Focus: Interactions between small groups of people. Topics: How groups work together, communication, team dynamics, leadership influence, and conflict. Examples: Addressing issues like communication inefficiencies. Conflict or team restructuring.
Organizational level	Focus: The organization. Topics: The overall structure, culture, strategy, and how the company interacts with its environment. Examples: Mergers, acquisitions, downsizing or major changes in systems or processes such as Lean or computer systems.

One of the most beliefs that most impacted management's decision was Belief #7, "Our prejudgments and biases can prevent us from finding the truth in what we see and hear." Management's prejudgments and their belief that people needed to be tough prevented them from having empathy. Their worldview did not allow for acknowledging vulnerabilities.

Social systems have levels that can be viewed in a few ways. In organization development, three levels are typically referred to as the individual level, group or team level, and organizational level. These levels focus on how changes, behaviors, and dynamics manifest at each of these distinct groupings within a company (see Table 5.2).

Imagine three concentric circles. The smallest circle at the center represents the individual, while the larger circle surrounding it is their immediate team or group within the workplace, such as engineering, sales, or mechanics. Each of these groups develops its own set of norms that influence the decisions and behaviors of that group. In fact, if they violate the norms, they are likely to be ostracized and not considered part of the team. The third, outermost circle represents the whole organization. This layer contains the organizational culture, systems, and hierarchy, which influence the beliefs and assumptions that differentiate one organization from another. It is also where the highest seat of power resides and the strategic direction is established.

The external environment, such as the economic environment, also exerts a powerful influence on behavior. The executive's role, therefore, is to navigate the external conditions from listening and understanding. This example is meant to illustrate how the intangible emotions and feelings of the workplace manifest depending on how employees feel treated. Are they trusted and valued, or are they considered unfeeling assets?

Analysis with Relationship-Centered Safety (RCS)

In the proactive RCS approach, the eight beliefs form the mental models that guide behaviors and decisions and the way people see reality. Figure 5.2 places these concepts in the first layer of the STS Iceberg.

Explanation of the STS Iceberg and RCS in Action

The first ergonomic injury triggers an inquiry into the patterns or trends that led to it. The supervisor brought up the problems caused by the reorganization of the warehouse. In a relationship-centered environment, all employees are heard when they bring up a safety concern. Therefore, an ergonomics assessment was ordered, and the employees met to evaluate the recommendations and how to implement them.

The mental models set the stage for the results. The eight beliefs that come forth in this scenario include #8, "Our pre-judgments and biases can prevent us from finding the truth in what we see and hear." The manager's pre-judgment ("if you get hurt, you are dumb") prevented him from seeing the ergonomic truth presented by the supervisor.

Belief#4, "People are willing and able to contribute to the success of the organization." If you hold this belief, you are willing to listen to concerns rather than disregard them as frivolous. This would have to be in place for management to listen and for the supervisor to propose a solution. In the first scenario, the supervisor took the risk to speak up, but it didn't do any good because management did not trust the workers.

These mental models (beliefs) are at the lowest level of the iceberg because they are usually uncouncious, unspoken, and taken for granted. In the original scenario (Figure 5.1), employees know they must take care of themselves because there is no respect for their contributions. Therefore, the company can encounter more workers' compensation insurance claims, absenteeism, and turnover. In the second scenario (Figure 5.2), you have an organization where people trust each other and have the open communication that bolsters the organization's ability to catch warning signs before they turn into injuries or fatalities.

The Truth Is Messy

Serious injuries and fatalities (SIFs) do not fit into any linear model of cause and effect. However, the iceberg helps to explain it as a process affected by hidden layers of mistaken assumptions, damaged relationships, goal-conflicts, and miscommunication. It illustrates why policies and systems can create a false sense of security—they simply don't affect the bottom layers. The iceberg bridges this disconnect between the "administrative world of systems and procedures" and the "real world of how the work is done" justifying the focus on the invisible layers (Mental Models and Relationships).

Shifting to a more collective/social approach won't be easy. The idea of individual accountability is deeply rooted in an ideology that prioritizes individualism and tangible over intangible evidence. This often leads to overlooking the intangible influence of peer pressure, stress, or personal emotional factors, such as fear of losing your job in a bad economy. Consequently, an unfair burden is placed on the individual who made the mistake, while the role of power dynamics, cultural, and even broader societal factors in defining acceptable risk go unseen.

Understanding How Social Systems Influence Safety

Recognizing and understanding the sometimes subtle behaviors and body language are as important as hazard recognition for the safety advisor. The workplace is a socio-technical environment (Rasmussen, 1997). The social part encompasses the intricate human interactions, emotions, and cultural dynamics that markedly shape safety outcomes.

A social system refers to an interconnected network of relationships and performance shaping factors. Every individual who belongs to this group shares cultural norms that influence their behavior, communication, and shares a common way of seeing the world. Everyone is part of an interconnected web, and a change in one person or team can affect others. This all shapes how decisions are made and how people are treated. Consequently the social system you belong to also shapes your view of reality.

Johnson's (1997) path of least resistance is reflected in each level of the organization. Each level includes sets of rules or social norms that entirely influence decision-making. He emphasizes that "the path" isn't just about individual choices but about the inherent dynamics within a system, which can guide people toward certain behaviors without them realizing it. This path favors conformity and maintaining the status quo, even when it means overlooking potential hazards or engaging in unsafe practices. This is borne out in industries located in disadvantaged neighborhoods with lower pay. The socioeconomic systems contribute to higher fatality rates because people feel they must take higher risks to make a living (CDC, 2024).

Which Level of the Organization Can Occupational Health and Safety (OHS) Best Influence

OHS can learn from other disciplines that also erroneously focus on the individual. Vivienne Cree (2015) challenged the concept of individual responsibility in her social work framework. She proposed a nuanced, collective, multi-layered perspective on responsibility, encompassing individual, familial, and social environments. Cree's exploration of collective responsibility led her to realize that individual human rights are contingent upon a shared sense of responsibility for one another and the environment. She explored how societal structures, family pressures, and organizational culture (all social constructs) can impact an individual's ability and willingness to change. This broader view highlights that an individual's actions and willingness to change are deeply influenced by these interconnected social factors, making singular individual blame unrealistic.

Cree's work draws a compelling parallel between social work and workplace safety, asserting that both fields focus on enhancing social environments to better address and safeguard well-being. Social work strives to create supportive environments for individuals and families, while safety science aims to design and implement systems that minimize harm. Cree's framework resonates with OHS practitioners who identify their work as part of the helping profession (Carrillo, 2022).

This further supports a shift from a compliance only model to a more sup portive, human-centered approach focused on collective well-being and systemic improvements, consistent with the ethic of care which will be further discussed in Chapter 9 as part of the strategic framework for RCS. This directly links the social environment—a collective construct—to the achievement of safety, moving away from a technical or individually focused approach (Leong et al., 2022).

This focus on the collective and systemic nature of safety naturally leads to an examination of the social processes that shape these dynamics.

Language as a Social Dynamic

The study of social dynamics examines how societies evolve, how trends spread, and how power structures shift. For example, virtual platforms changed the way we communicate. In a similar manner, language is an integral part of an organization's social dynamics that changes how safety culture evolves and shifts in power, for example, shifting from top-down control to a decentralized framework. Language itself is a continuous and dynamic social process (Mallinson, 2025).

This process takes place as the meanings of words and symbols are defined through daily conversation and discourse. These shared meanings form the basis of our common knowledge and social norms. For example, the way we talk about something, like mental health, can either destigmatize it or reinforce negative stereotypes, demonstrating the power of language to shape social reality.

Because language shapes reality, it is also a source of power. Those in power often define what language we use. Those who have power define what is considered "normal," "true," or "important." Language can reinforce existing power dynamics, but it can also be a tool for challenging them, as seen in social justice movements that aim to create alternative perspectives.

From a sociolinguistic perspective, language, culture, and society influence each other in ways that are inseparable. These perspectives indicate that to achieve a significant reduction in SIFs, a paradigm shift is needed in the language we use, from "I am responsible for safety" to "we are responsible." This shift involves recognizing the interconnectedness of workers and the importance of social and psychological factors in safety (Long, 2019).

Leaders play a vital role in driving this shift by creating a workplace that values relationships, open communication, and worker's experience. They also ensure that workers receive the training and resources they need to identify and address risk. The next topic, the application of non-technical skills (NTS), is one of those resources.

What NTS Contribute to the Ways We Interact with Risk

Following the Deepwater Horizon event, the offshore oil and gas industry became interested in applying Crew Resource Management NTS to enhance safety (Flin & Agnew, 2014). Social scientists' analyses of the accident clearly show how failures in NTS contributed to the trajectory of that event (Hopkins, 2012; 2023; Reader &O'Connor, 2014; Roberts et al., 2015). They found that effective accident prevention is deeply intertwined with how we use language to construct shared understanding, build relationships, and navigate the social and emotional landscape of the workplace.

NTS are the essential social and cognitive competencies required to navigate complex systems. Now, we will examine how these skills function—or fail to function—within the socio-technical container of the modern workplace, where systemic pressures often create a "path of least resistance" toward unsafe acts.
NTS could also be called social competence skills. This is a critical area for leadership development because a leader's ability to influence relies on these skills. It is also the weakest area for many managers because the importance of language has been poorly understood.

> Two aspects of social competence can be highlighted as being especially relevant for senior managerial influence on safety: social perception and persuasion [...] emphasise the ability to understand the emotions of others as being among the strongest predictors of senior managers' effectiveness and this has also been highlighted for the influence on safety. A senior manager's ability to show active involvement and communicate effectively will be facilitated by their social perception skills.
>
> (Fruhen et al., 2014)

NTS is about task completion in a safe and efficient manner and are meant to improve communication, trust, and mutual support. - The ability to perceive the emotions of others—a key NTS aspect of social competence—is one of the strongest predictors of a senior manager's effectiveness and influence on safety. The failure of management in the warehouse to exhibit these skills is a clear example of a leadership failure, not just a system failure. Sometimes these communication barriers are not conscious, so again the leader must be trained on how to recognize the signs.

NTS Failure during Fukushima Disaster

According to the researchers of the Fukushima disaster, the absence of these same NTS led to the failure of the defense systems. A major factor in the Fukushima disaster was the widespread assumption that nuclear power plants were so safe that an accident of this magnitude was "unthinkable." This belief led operators, regulators, and the government to accept that the Fukushima plant could handle any external event, even though safety standards and tsunami defenses lagged international best practices. People close to these events had done something deeply human; they had normalized the risks associated with their work. Pupulidy & Vesel, (2023) describe this in detail:

> Normalization of risk is a common aspect of long-term human interactions with risk, regardless of any calculation of risk or risk management process. The review of accidents showed that the longer a system seemed safe (no accidents), the greater the amount of risk people unknowingly accepted.

In such an environment, NTS training is like giving a crew life vests but locking them in a cabin as the ship sinks. NTS are critically vulnerable to the social context. You can argue that the greatest "unseen risk" in an organization is not a lack of skills but a culture that renders those skills useless. As a report after the Fukushima disaster advises, if we identify and extract in-depth causes behind organizational failures, and cultural factors are suggested to exist, this can be used to establish risk and crisis management that considers cultural aspects such as groupism, obedience to authority, willful blindness, culture of silence, and immaturity of just culture (Gil, 2021).

This is often confused with a more frequently used word "complacency." Normalization of Risk allows for more understanding and learning to take place and helps us to understand how the path of least resistance can be accepted by quality frontline workers without recognizing what was happening. Individuals need strong situational awareness and psychological safety skills to recognize when the "easy path" is also the unsafe one. Creating a culture where challenging that path is supported is difficult, even with the best training in the world (Bauer et al., 2018).

However, these skills are not exercised in a vacuum. They are either enabled or crippled by the organization's social system. A post-Fukushima disaster report offered a chilling warning about the danger of creating an organization "where open discussion is impossible." In such environments, where social pressures create a powerful 'path of least resistance' toward silence, NTS are not a solution; they are a casualty (Murata, 2021).

This is not just a nuclear industry problem; it is a human systems problem that applies to any high-pressure, hierarchical environment where conformity is valued over curiosity. It makes it very difficult for people to speak their truth, even the best-trained teams are set up to fail. The true challenge for leadership, therefore, is not simply to train for NTS but to build a culture of trust and psychological safety where those skills can be used.

The failure of these essential human competencies within a high-stakes STS was a central cause of the Deepwater Horizon disaster, a tragedy we will deconstruct in detail in Chapter 10. These NTS are not optional "soft" abilities; they are the core of a system's resilience (IOGP, 2014). The impediments to transparent communication are real and unassailable unless a leader can influence the creation of open communication systems, as will be discussed in Chapter 9.

See more on "Core Non-Technical Social Skills" in Appendix C.

STS as an Organizational Framework

Awareness of social systems is important in high-risk industries because many risks are in the social systems which tend to be ignored in OHS.

The dominance of individualism in Western thought, influenced by the scientific method's emphasis on individual observation, isn't new and presents significant challenges for leading safety. While a sense of individual responsibility can spark change, nothing changes unless there is a following. That is the essential role of social context and group dynamics in shaping the actions and choices of individuals within a group setting (Thibaut & Kelley, 1959). Chapter 4 covers the origins of STS, but here it serves as a vision of what this book is moving toward—a view of effective organizations that recognizes the strengths and weaknesses of both people and technology.

Individualism ignores the powerful influence of systemic organizational factors on incidents. STS, in contrast to the individualistic view that "accidents are the result of human error," forces attention onto deeper organizational issues. And normalized unsafe procedures and practical realities (like immediate task demands) that collectively make conforming to unsafe practices appear far simpler and less risky than actively challenging them.

This leads to a system where employee well-being is not prioritized, resulting in high stress levels, burnout, and persistent safety issues. A common misconception arises here, where some believe that emphasizing systems absolves people of responsibility. However, this is a misunderstanding. People are not separate from these systems; they are integral components, actively shaping and being shaped by them. The reality is that we operate within socio-technical containers, where human actions, decisions, processes, and technology intertwine to produce organizational outcomes.

An STS involves understanding information flow, decision-making processes, communication patterns, and how different parts of the system contribute to the overall function. This functional perspective makes it possible to move toward recognizing the powerful influence of social processes (Grimaud, 1991). Traditional approaches tend to down-play or devalue the interconnectedness of people within the system and the ways in which people affect each other. Yet, research continues to demonstrate worker's abilities to reach higher levels of performance via collaboration with each other.

Why Would Self-Organizing Workforces Have Fewer Accidents?

Research by Dwyer and Raftery (1991) confirm the STS legacy of Emery and Trist's research in the mines of the UK (Chapter 4). They provided evidence from worksites showing that where the workers had greater control led to a reduction in accidents. Specifically, on night shifts, the decreased presence and pressure from management fostered an environment where workers could exercise "auto-command." This empowerment enabled them to refuse unsafe work or challenge management's authoritarian approach, resulting in lower accident rates.

During these less-managed night shifts, workers developed informal self-regulatory practices to manage risks. For instance, they allowed overly tired colleagues to take unauthorized breaks, a practice forbidden during the day, which helped mitigate fatigue-related hazards. They also proactively self-organized by rotating routine tasks and correcting disorganization themselves, minimizing the negative influence of organizational issues on accident rates.

These qualitative findings consistently demonstrate a strong correlation: transforming social relations, particularly by granting workers more autonomy, directly leads to changes in accident rates. The night shift's accident rate was dramatically lower, showing a 93% reduction.

Their evidence is blunt: when you give workers power, they stop getting hurt. The night shift data provided proof that safety doesn't come from more oversight; it comes from more autonomy.

Power as a Social System

From this viewpoint, a safety management system is not a neutral, technical framework; it is a manifestation of power that produces a particular version of reality, defining what is knowable, sayable, and doable with respect to safety. In the social sciences, power is a social system. Instead of being a quality an individual possesses, power is viewed as embedded in social structures, relationships, beliefs, and practices that influence and control people's behavior and access to resources.

An examination of landmark workplace disasters across different industries reveals that although the industrial contexts have changed, the power and corporate dynamics leading to tragedy have remained remarkably consistent. Egregious industrial disasters are not isolated failures but are manifestations of a shared and persistent pattern of prioritizing production and profit over worker safety.

For example, on April 5, 2010, an explosion at Massey Energy's Upper Big Branch mine in West Virginia killed 29 of the 31 (McAteer, 2011) coal miners at the site. Chief executive officer (CEO) Dan Blankenship had received daily updates on safety violations and yet had a history of encouraging mine supervisors to focus on "running coal" rather than complying with safety standards. Blankenship was sentenced for conspiracy to willfully violate mine health and safety standards after a jury returned a guilty verdict on the federal crime. He was sentenced to a year in federal prison and ordered to pay a $250,000 fine (McAteer, 2011).

This incident represents a toxic form of excessive individualism where one person's ambition and disregard for the workforce motivated a willful violation of safety standards and human values. Multiple disasters discussed in this book followed this pattern including Deep Water Horizon and Bhopal.

Crafting a Safe Workplace through Collective Responsibility

Albert Einstein, celebrated for his scientific brilliance, was also a fervent advocate for social justice, a radical stance often obscured by his towering legacy in physics. His experiences, from escaping Nazi Germany to observing the fervent individualism of the United States, cemented his belief that societal well-being hinges on prioritizing our collective nature. Einstein recognized that although individuals possess both social and individualistic drives, the capitalist societies of his time, much like ours, relentlessly accentuate the latter, suppressing the very bonds that foster human thriving and, crucially, safety.

The prevailing ideology in many Western societies, particularly the United States, champions an individualism that narrowly defines responsibility, often limiting it to one's immediate circle. This worldview fosters a competitive struggle for survival, where "grabbing as much as you can" is presented as the path to security. Yet, as Einstein (2025 Ed.) passionately articulated in "Why Socialism," this economic anarchy is "the real source of evil." It pits individuals against each other, incentivizing them to "deprive each other of the fruits of their collective labor," even when done through "legally established rules."

Human beings are not built to flourish in isolation. Our social nature is vessential to our well-being and survival. The suppression of this inherent social drive, as Einstein warned, leads to a profound "deterioration" of the human spirit, leaving individuals "insecure, lonely, and deprived of the naïve, simple and unsophisticated enjoyment of life." This is not merely a philosophical concern; it has tangible, devastating consequences for our collective safety and well-being.

Consider the profound implications of shifting our understanding of safety from "I am responsible for safety" to "We are responsible for safety." The former, although seemingly empowering, often leads to a false sense of control and can foster a dangerous fear or shame of reporting errors, knowing one will be in a negative spotlight. "We are responsible" however, demands trust—a willingness to depend on others to recognize our inescapable interconnectedness. While some may shutter at the thought of depending on others, this aversion often stems from an unconscious denial of our fundamental reliance on one another. We cannot escape our interconnectedness, "for better or for worse."

While acknowledging that safety statistics are not solely a product of cultural orientation, it is instructive to observe nations with robust social support programs. Countries like Norway, Denmark, and Sweden have some of the lowest fatality rates in the world. This suggests that while technical advancements are required for safety, they are amplified and made more effective when embedded within a societal structure that prioritizes collective well-being and mutual responsibility (Lu, Jin & English, 2021). The investment in "the technical side of keeping people safe" is more likely to occur and be equitably distributed in a society that values the lives and safety of all its members, not just those who can "grab" enough resources.

Social justice demands a deep re-evaluation of our societal values. It calls for us to move beyond the stifling confines of radical individualism and embrace the collective responsibilities that define our humanity. By fostering a society where our social drives are accentuated and supported, we can cultivate genuine safety, belonging, and the equitable distribution of resources, enabling everyone to thrive. As Einstein understood, "Man can find meaning in life, short and perilous as it is, only through devoting himself to society." This is not just a call for empathy; it is a blueprint for a more just, secure, and ultimately, more humane future.

Enhancing Collective Responsibility

A sense of collective responsibility can be enhanced by the following safety practices as outlined by Yorio et al. (2019) in their comprehensive analysis of Safety Culture across Cultures. These strategies work best in a collective culture because they are aligned with the core values and social dynamics of those cultures. They don't try to change the culture; they leverage it.

Reinforcing Group Identity: In collectivist cultures, an individual's identity is often defined by their membership in a group. Therefore, structures like team-based work and group incentives naturally appeal to a person's sense of purpose and belonging. The shared success or failure of the group becomes a more powerful motivator than individual rewards.

Leveraging Social Norms and Trust: Collective cultures have a strong emphasis on social harmony and shared responsibility. When you use group safety training or mentoring programs, you are tapping into the existing networks of trust and peer influence. Workers are more likely to listen to and learn from their peers, who are part of their in-group, than from a top-down directive.

Aligning Goals: For these cultures, the organization's goals are seen as a communal effort. Efforts to facilitate cooperation are interpreted as ways to support the group's well-being and success. This means that a team goal-setting approach is more effective because it merges individual goals with the group's goals, creating a powerful sense of shared purpose for safety.

These same strategies often fail in an individualistic culture because they directly conflict with core cultural values such as personal autonomy, individual achievement, and self-reliance. Instead of being motivating, they can lead to resentment and a lack of buy-in. These same values often make team incentives and recognition ineffective.

Discomfort with Peer Pressure: The emphasis on personal autonomy means individuals often resent peer-based monitoring or pressure, viewing it as an intrusion on their freedom. A peer-to-peer mentoring program might be perceived as an unwanted intrusion rather than a helpful tool.

Moving toward Collectivism in Safety

There is an urgent case to be made to move toward collectivism now. As individualism becomes more culturally dominant globally, people are increasingly expected to self-regulate, self-advocate, and "own" their well-being, even when emotionally overwhelmed. This is reflected in workplace norms that prioritize self-reliance over collective support. We can see these effects in the rising mental health crisis revealed in global surveys.

- Twelve billion working days are lost every year due to depression and anxiety according to a joint WHO/ILO (2022) study.
- In the United Kingdom alone, 17.1 million workdays were lost in 2022/23 due to work-related stress, anxiety, and depression (49% of all days lost due to ill health) according to data from the UK Health and Safety Executive (Evotix, 2023).
- Nearly two-thirds (65%) stated that work adversely affects stress levels in ADP's People at Work 2023 Survey, an enormous if slightly reduced number versus 2022 (Richardson & Antonello, 2023).
- Forty-five percent of employees say poor mental health is having an adverse impact on productivity in Asia according to the Asia Mental Health Index, a recent study from AON and Telus Health (ERM, 2024).

The 256 Health and Safety Function Leaders who participated in the ERM study inform us that the costs for organizations that fail to rise to the challenge will continue to increase for the foreseeable future. For example, concerns expressed by OHS practitioners that they are not educated to deal with psychosocial or mental health issues will have to be addressed by the educational system and regulatory bodies.

It is inevitable for the social psychological aspects of the workplace to become an employer responsibility. There is no doubt that workplace stressors, such as excessive workload, harassment, or lack of support, are harmful to emotional health, which in turn can increase the risk of accidents (John Hopkins University, 2023). The only question is Where will that responsibility lie? If it lands in OHS, there will have to be an enormous shift in OHS education.

The Shift to Collectivism

While employers hold the primary duty to generate conditions for collaboration, trust, and open communication, the buy-in of every team member is crucial.

In any complex STS, preventing catastrophic incidents demand a substantial shift in perspective. A truly effective approach recognizes the influence of broader social systems and the intricate dynamics within an organization. This means understanding how organizational culture, communication patterns, and even external societal pressures can shape individual behaviors and systemic vulnerabilities

Furthermore, a truly safe environment acknowledges the inherent interdependence of team members. While individual contributions are undoubtedly valuable, lasting safety improvements are intrinsically group-driven outcomes. They necessitate collaboration across all organizational levels. When teams operate with a shared understanding of risks, clear communication channels, and a clear understanding of roles, the entire system becomes more resilient to unexpected challenges. Collective intelligence and coordinated effort are far more powerful than the sum of individual actions in preventing incidents and mitigating their impact.

Steven Sloman and Philip Fernbach (2017) argue for collective rather than individual intelligence in "The Knowledge Illusion" that individual human intelligence is surprisingly shallow. We think we understand how things work, but we rely on a "community of knowledge." This supports the idea that no individual (pilot, surgeon, CEO) is ever truly "autonomous," yet our liability and reward systems presume they are. Unfortunately, as Geoff Mulgan argues in *Big Mind*, organizations succeed due to "Collective Intelligence" (the orchestration of many minds/machines), yet our management structures still have an obsession with the "hierarchy of individuals."

To cultivate such a collective environment, leadership must actively role-model transparency, readily admit mistakes, and become active learners. This act of vulnerability is not a sign of weakness but is, in fact, key influence for employees to engage in open communication without fear. When leaders demonstrate a genuine willingness to acknowledge their own errors, whether in decision-making, resource allocation, or process design, it signals to the entire workforce that reporting concerns, near misses, or actual errors will not be met with punishment or humiliation.

This isn't just theory. Transparency and vulnerability came up as admired leadership characteristics in Google's research with their high performance teams. Inspired by that report, I searched for a leader on Linked-In that was respected and appreciated by their staff. I received several recommendations to interview Adrienne Kelbie, who was at the time CEO for Nuclear Regulation UK. So I interviewed her staff with the same questions used by Google. Her direct reports said that one of her most outstanding characteristics was her willingness to apologize if she made an error and help others learn from their mistakes. She had regular one-to-one conversations with each where they discussed their career development. In such an environment, each member of the organization becomes a vital sensor for early warnings that can prevent failures. This distributed intelligence allows for proactive problem-solving and adaptive responses to emergent risks.

Beyond reactive responses, organizations must actively engage in open dialogues about difficult decisions and potential trade-offs before an incident happens. This proactive communication ensures that people know where the "guard rails" are—the acceptable boundaries and safety margins. It involves a delicate balance: treating mistakes as invaluable opportunities to learn while simultaneously holding individuals accountable for gross negligence. This approach fosters a learning environment, not a punitive one, and is essential for continuous improvement in safety performance.

Conclusion

This chapter establishes a decisive link between organizational safety and the social dynamics of its operational systems—specifically, the often-hidden interactions between people, technology, and organizational structure. To illuminate these underlying forces, the STS Iceberg is introduced as an essential diagnostic tool designed to expose these systemic vulnerabilities.

Top-down safety management is essentially flawed and unsustainable. New protocols fail without a strong organizational foundation built on deep trust and a shared understanding of risk. This means leadership must shift from enforcing compliance (which fosters blame) to empowering frontline teams. The mandate is to enable them to actively manage risk, solve problems locally, and build safety practices organically. In this state, safety is not a reactive add-on but an emergent property—a natural, self-sustaining outcome of a well-functioning, high-trust STS.

Executive Takeaways

- Abandon the ideology of individual responsibility as the primary driver of safety, recognizing it as a flawed myth that obscures organizational problems and protects the power structure.
- Train safety facilitators, supervisors, managers to conduct safety incident reviews as systemic failures rooted in the complex STS, rather than the simple sum of individual actions.
- Provide leadership development that focuses on social and problem-solving skills, as empirical evidence shows these, not technical safety knowledge, demonstrate a senior manager's commitment to safety.
- Specifically target the development of social perception—the ability to understand the emotions and intentions of others—as the most critical aspect of social competence for executives seeking to influence safety.
- Review of production goals and organizational policies to identify and eliminate decisions that inadvertently create a "path of least resistance" toward unsafe behaviors for employees.

References

Bauer, M. W., Gylstorff, S., Madsen, E. B., & Mejlgaard, N. (2018). The Fukushima accident and public perceptions about nuclear power around the globe –A challenge & response model. *Environmental Communication*, 13(4), 505–526. https://doi.org/10.1080/17524032.2018.1462225

Carrillo R. A. (2021, 2024). Unpublished report.

Carrillo, R. A. (2021). Personal conversation.

Carrillo, R. A. (2022). *Safety Leadership Strategy*. Routledge.

CDC. (2024). Socioeconomic Factors at the Intersection of Race and Ethnicity. www.cdc.gov/disability-and-health/articles-documents/socioeconomic-factors-race-and-ethnicity.html#:~:text=At%20a%20glance,adults%20(1%20in%2010).

Cree, V. (2015). Responsibility in social work. InWright, J. D. (Ed.), *International Encyclopedia of the Social & Behavioral Sciences* (2nd ed.). Elsevier, 581–586.

Dwyer, T., & Raferty, A. E. (1991). Industrial accidents are produced by social relations of work: A sociological theory of industrial accidents. *Applied Ergonomics*, 22(3): 167–178. https://doi.org/10.1016/0003-6870(91)90156-C

Einstein, A. (Ed). (2025). Albert Einstein's "why socialism?": The enduring legacy of his classic essay. *Monthly Review Press*: NY.

ERM. (2024). Global Health and Safety Survey. www.erm.com/globalassets/insights/erm-2024-global-health-and-safety-survey-final-report.pdf

Evotix. (2023). A Breakdown of HSE's Key Health & Safety Figures for Great Britain (2022/23). www.evotix.com/resources/blog/a-breakdown-of-hses-key-health-safety-figures-for-great-britain-2022/2

Flin, R., Wilkinson, J., & Agnew, C. (2014). *Wells Operations Crew Resource Management. International Oil and Gas Producers Report 501*. London: IOGP.

Fruhen, L. S., Mearns, K. J., Flin, R., & Kirwan, B. (2014). Skills, knowledge and senior managers' demonstrations of safety commitment. *Safety Science*, 69: 29–36.

Gil, L. (2021). *Learning from Fukushima Daiichi: Factors Leading to the Accident*. IAEA. www.iaea.org/bulletin/learning-from-fukushima-daiichi-factors-leading-to-the-accident

Grimaud, M. (1991). Psychology and the Literary Study of Literature. Poetics, 20(3): 221–259.

Hopkins, A. (2012). *Disastrous Decisions: The Human and Organisational Causes of the Gulf of Mexico Blowout*. CCH Australia.

Hopkins, A. (2023). *The Human and Organisational Causes of the Gulf of Mexico Blowout*. Australian National University. Elgar Concise Encyclopedia of Oil and Gas Law.

ILO. (2022).Nearly 3 million people die of work related accidents and diseases. www.ilo.org/resource/news/nearly-3-million-people-die-work-related-accidents-and-diseases#:~:text=The%20Asia%20and%20the%20Pacific,agricultural%20workers%2C%20the%20report%20says

International Association of Oil and Gas Producers, IOGP. (2014). *Human factors 101*. https://humanfactors101.com/topics/non-technical-skills-crm/

John Hopkins University. (2023). Why should mental health be a priority in the workplace? https://imagine.jhu.edu/blog/2023/05/17/breaking-the-stigma-why-mental-health-should-be-a-priority-in-the-workplace/#:~:text=Stress%20and%20burnout%20are%20the,or%20instruction%20for%20the%20role?>

Johnson, A. G. (1997).*The Gender Knot: Unraveling Our Patriarchal Legacy*. Temple Diversity Press.

Leong, S., Eom, K., Ishii, K., Aichberger, M. C., Fetz, K., Müller, T. S., et al. (2022). Individual costs and community benefits: Collectivism and individuals' compliance with public health interventions. *PLoS ONE*, 17(11). https://doi.org/10.1371/journal.pone.0275388

Long, R. (2019). *The Social Psychology of Risk Handbook*. Scotoma Press.

Lu, J. G., Jin, P., & English, A. S. (2021). Collectivism predicts mask usage during the COVID-19 pandemic. *Proceedings of the National Academy of Sciences*, 118 (23).
Mallinson, K. (2025). *Sociolinguistics*. Britannica. www.britannica.com/science/sociolinguistics
McAteer, J. D. (2011). Upper Big Branch: The April 5, 2010, Explosion: A Report to the Governor of West Virginia. West Virginia Governor's Independent Investigation Panel.
Mulgan, G. (2017). *Big Mind: How Collective Intelligence Can Change Our World*. Princeton University Press.
Murata, A. (2021). Advances in Physical, Social & Occupational Ergonomics: Proceedings of the AHFE 2021 Virtual Conferences. Eds. Goonetilleke, R., Xiong, S., Kalkis, H., Roja, Z., Karwowki, W. https://link.springer.com/book/10.1007/978-3-030-80713-9#author-1-0.
Partheeswaran, P., & Shaf, M. S. (2025). Criminological and socioeconomic aspects of corporate delinquency: A 21st century perspective. *Journal of Economic Criminology*, 9: 100185. https://doi.org/10.1016/j.jeconc.2025.100185
Pupulidy, I., & Vesel, C. (2023). *Human & Organization Potential*. Dynamic Inquiry LLC.
Rasmussen, J. (1997). Risk management in a dynamic society: A modelling problem. *Safety Science*, 27(2–3): 183–213.
Reader, T., & O'Connor, P. (2014). The Deepwater Horizon explosion: nontechnical skills, safety culture, and system complexity. *Journal of Risk Research*, 17: 405–424.
Richardson, N., & Antonello, M. (2023). People at work 2023: A global workforce view. www.adpresearch.com/wp-content/uploads/2023/04/People-at-Work-2023_A-Global-Workforce-View.pdf
Roberts, R., Flin, R., & Cleland, J. (2015). 'Everything was fine' An analysis of the drill crew's situation awareness on Deepwater Horizon. *Journal of Loss Prevention in the Process Industries*, 38: 87–100.
Rochlin, G. I. (1999). Safe operation as a social construct. *Ergonomics*,42(11): 1549–1560.
Sloman, S., & Fernbach, P. (2017). *The Knowledge Illusion: Why We Never Think Alone*. Riverhead Books.
Srivastava, M., & Mishra, T. (2025). The economic impact of white collar crime on India's growth.*International Journal of Advanced Research in Scienc*e, 5(1): 605–609.
Thibaut, J. W., & Kelley, H. H. (1959). *The Social Psychology of Groups*. Wiley.
Weick, K. (1988). Enacted sensemaking in crisis situations. http://wendynorris.com/wp-content/uploads/2018/08/Weick-1988-Enacted-Sensemaking-in-Crisis-Situations.pdf
Weick, K. (2024). Mann gulch revisited: Improvisation as a surface of apprehension. *Academy of Management*. https://doi.org/10.5465/amd.2024.0006
Yorio, P. L., Edwards, J., & Hoeneveld, D. (2019). Safety culture across cultures. *Safety Science*, 120: 402–410. https://doi.org/10.1016/j.ssci.2019.07.021

6 Risk as a Social Construction

Introduction

Risk assessment managers will need to learn about cultural and social aspects of risk analysis, management, and communication. Most theories of risk do not attempt to understand the human dimension. The express goal of this chapter is to bring together, psychological, social and cultural theories. This realization is part of a broader societal shift away from treating risk solely as an engineering problem. Prophetically in 1992, Collins, a noted risk expert warned,

> People are disbelieving hard scientists. Local, state, federal, and international policy makers are being asked to make hard risk based decisions about soft science, as opposed to the earlier practice of the fading technocratic paradigm of making soft, value laden and compromised decisions based on hard science.
> (1992: 285)

That time is now.

Mary Douglas (1992) sets the stage for this conversation. She revolutionized the discipline of industrial analysis insisting that risk is a social and relational process rather than purely an individual or psychological one. In her book, *Risk and Blame*, she argues that risk is a social and cultural issue. An example she gives is that experts are constantly baffled that people don't buy earthquake insurance or prepare for obvious danger. Douglas argues the reason experts are so confused is that they are committed to studying risk by looking at one person at a time, trying to find a purely logical or "objective" reason for their decisions.

She wrote that the mistake is thinking that risk decisions are made objectively—they aren't, not even trivial ones. Emotions are part of the riskiest decisions, motivating people to mitigate risk by not making important decisions without consulting neighbors, family, work, or friends. Yet, the risk perception research says practically nothing about intersubjectivity, consensus, or the social influences on decisions. By focusing only on individuals, researchers miss the most important influential part of the puzzle: our relationships and social connections.

Mary Douglas concludes with her core idea. She believes that human beings are social creatures first and foremost. She uses the metaphor of *control gates to*

DOI: 10.4324/9781003668770-6

mitigate risk. Our social life—our need to belong, our loyalty to our groups, our desire to protect our way of life—acts like a gatekeeper for information. We don't just accept facts as facts. We only accept information that the groups we belong to believe. Or that comes from another source we trust. If it comes from an "outside" group, the gate slams shut. We ignore it, dismiss it, or find a reason to call it false. Douglas was among the first to recognize blaming as a form of self defense.

> Blaming is a way of manning the gates through which all information has to pass. Blaming is a way of manning the gates and at the same time of arming the guard. News that is going to be accepted as true information has to be wearing a badge of loyalty to the particular political regime, which a person supports: the rest is suspect, deliberately, censored, or unconsciously ignored from this standpoint, the proper way to organize the program of studying risk is to start with studying institutional design.
>
> (Douglas, 1992: 12)

Therefore, if you want to understand how people perceive and act on risk, you must stop studying individual behavior and widen your understanding of how risk is perceived and determined.

Workplace Safety Is a Social Construct

Safety and risk are frequently considered objective truths. However, this view oversimplifies both, reducing them to a cause-and-effect mechanism. In this framework, accidents are attributed to either a "breakdown" or "human error." This is a view largely separate from social influences (Nichols, 1999; Shufutinsky et al., 2017).

Yet, risk isn't simply a factual measure of potential danger. Instead, it is a social construct influenced and shaped by various social elements, including cultural norms, power relationships, and how we communicate about it (Turner & Gray, 2009; Nichols, 1975, 1999; Chikudate, 2009).

Prominent researchers in the safety field have explored this understanding of safety and risk as constructs (Rochlin, 1999; Taylor, 1981; Tierney, 1999). These researchers have examined how social processes and culture influence organizational safety outcomes, especially in the context of disasters and serious injuries and fatalities (SIFs). Tetteh (2025), recounts a workers' incapacitating accident that illustrates how the social construct of power between a supervisor and his direct report led to tragedy. It involves power relations and production pressures, often the social constraints that lead to SIFs.

> Mike, a diligent employee, was assigned to remove a 480-volt switch panel. The standard Lockout/Tagout (LOTO) procedure required him to personally verify that a breaker was pulled and apply his own lock. He was about to leave to do so when his supervisor told him not to worry because he had already applied his own lock, urging Mike to hurry and emphasizing the "minimal effort required."

> Despite an "uneasy feeling" and an internal voice telling him to verify, he proceeded, eager to help. As he opened the breaker box, all went black and he was severely shocked, resulting in permanent nerve damage and an irrevocably changed life.

The story constructs safety as an evolving target. It is why implementing a system or training protocol is insufficient. While formal regulations and designs set baselines, they have inherent limits and cannot fully address the variety and unpredictability of real-world work, thus creating a gap between intended compliance and actual safety. This leaves us with the reality that conversations in real-time ultimately determine risk decisions.

Analysis of Mike's Experience through the Socio-Technical Iceberg

Figure 6.1 illustrates how to use the ST Iceberg to reveal the hidden elements contributing to an accident.

Event (The Tip of the Iceberg)

What happened? Mike, an electrician, was severely shocked while removing a 480-volt switch panel, resulting in permanent nerve damage. This is the visible, acute incident that traditional safety management often focuses on investigating.

Patterns and Trends: What's been happening over time? The story describes a broader organizational culture that may prioritize efficiency or speed over

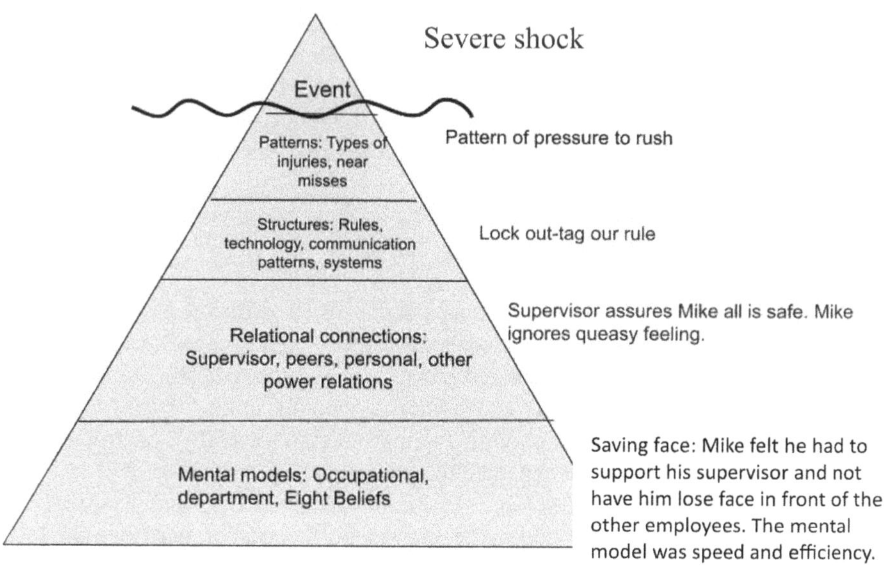

Figure 6.1 Iceberg analysis of Mike's accident.

strict adherence to safety protocols. The supervisor's language—urging Mike to "hurry" and emphasizing the "minimal effort required"—suggests a pattern of normalizing shortcuts and deviations from procedure in the name of production.

Underlying Structures: What are the systems and rules? There are two conflicting structures at play:

- *Formal Structure:* The official Lockout/Tagout (LOTO) procedure, which required Mike to personally verify the breaker and apply his own lock.
- *Informal Structure:* The unwritten rules of the workplace, which include the hierarchical power structure (the supervisor has authority) and the production pressure to get the job done quickly. At this moment, the informal structure overrode the formal one.

Mike's decision to proceed despite his "uneasy feeling" highlights the social construction of trust, particularly in hierarchical relationships. He trusts his supervisor's word, not just because he's an individual but because of the social role he occupies (supervisor = authority, knowledge, responsibility). This trust is a product of repeated social interactions and the established power structure within the workplace. If workers were consistently told to trust their lived experience, they might feel that questioning a supervisor is part of accepted protocol.

Relationships and Connections: How do we relate to others? The relationship is a hierarchical one between a supervisor and a direct report, but it is also one of mutual support. In this dynamic, the supervisor uses his authority to reconstruct the meaning of safety in real-time. "Trust me." Mike's trust in his supervisor is a product of this established power structure, but it can also be a reciprocal relationship where Mike wants to help the supervisor meet the production pressures.

This dynamic demonstrates a failure of RCS Belief #7: Our pre-judgments and biases can prevent us from finding the truth in what we see and hear. The assumptions about what constituted respect and support between Mike and his supervisor prevented true, open dialogue about the risk, silencing Mike's "uneasy feeling".

Mental Models: What do we believe? The incident was driven by a collision of several powerful mental models.

- *The Mental Model of Authority and Trust.* Mike's decision to proceed was heavily influenced by the supervisor's assurance, "don't worry because he had already applied his own lock," acted as a powerful social cue that overrode both the formal procedure and Mike's own internal warning signs ("uneasy feeling"). The trust wasn't a personal failing; it was a product of a socially constructed role where authority figures are expected to be correct.
- *The Mental Model of "Saving Face" and Fear of Social Risk.* There is a powerful human need to "save face" and avoid social risk. To insist on verifying the lock would have meant directly contradicting his superior, an act that could be seen as

disrespectful or challenging. The sources note that human interaction is a high-risk activity where people fear being embarrassed or humiliated. Challenging a boss in front of others is a significant interpersonal risk. Therefore, Mike was caught between a known physical risk (electrocution) and an immediate social risk (losing face, being seen as untrusting or insubordinate). At that moment, the social pressure was more real.

- *The Supervisor's Mental Model: Efficiency over Adherence.* The supervisor was operating from a mental model where production and efficiency are the primary goals. His language—"hurry" and "minimal effort"—frames the situation through this lens. He socially constructs a reality where the formal LOTO procedure is an unnecessary delay, and his personal assurance is a sufficient substitute. This creates a "path of least resistance" where following his directive seems easier and more acceptable than adhering to the rule.

Hopefully by now it is clear why training on procedures or even speaking did could not change the dynamics that drove this result. Mike's life changing injury was not a simple case of human error. It was the predictable outcome of a system where powerful mental models—trust in authority, the fear of social risk, and the prioritization of efficiency—collided and created a situation where a formal safety procedure was easily overridden by informal social dynamics. Mike's "uneasy feeling" was his individual risk assessment, but it was silenced by the powerful, socially constructed reality created by his supervisor and the workplace culture.

Sensemaking and Saving Face

Was Mike's situation something that psychological safety could have prevented? Human dynamics are far more complicated. "Human interaction has always been and remains the most frequent and riskiest activity in the workplace, and in life. ...," Goffman concludes, "There is no interaction in which the participants do not take an appreciable chance of being slightly embarrassed or a slight chance of being deeply humiliated. Life may not be much of a gamble, but interaction is" (Trevino, 2003:14).

Genuine organizational transformation requires a deeper engagement with the intricacies of human interaction and interpretation. Culture, as defined by Schein (1992), is a pattern of shared basic assumptions—a collective, ongoing "performance" that shapes how an organization presents itself ("front stage") and its deeply embedded, often unconscious, beliefs ("back stage"). Effective safety management requires bringing these "backstage" assumptions to light for re-evaluation. This is easier said than done, and will only reveal themselves through conversation.

There is a reason that such conversations are often neglected and there is a preference for written communication, or, more fundamentally, the inherent fear of human interaction itself. Yet, as personal anecdotes from safety advisors reveal, genuine communication—characterized by managers who care and are comfortable engaging with their teams—is highly valued by employees. This highlights the critical need for leaders to cultivate empathetic communication and a "language of care," recognizing the emotional labor inherent in leadership and advisory roles. The ability to go into the field, ask questions, and engage authentically provides access to accurate information about frontline realities that formal channels often miss, implicitly arguing that genuine communication and relationship-building can help overcome the "face-saving" instincts that undermine safety.

The fact that these dynamics are rarely if ever explored in OHS discussions is a testament to the threat they represent to all organizational members. Speaking truth to power or anyone is not to be taken lightly. I remember Ed Schein's advice to our organizational development class: never go deeper than the client can accept without losing face. It was a lesson I had to learn each time I got fired.

George Herbert Mead (2025) argued that we would not know we existed if we lived alone in a forest. It is through interactions with others that we receive feedback and judgments that shape our understanding of who we are. This has deep implications for how we understand the world and how people affect each other's decisions and actions. It could be argued that our human ego or identity relies on being seen and heard by others. Saving face isn't just about an individual's ego; it's a social process of maintaining a positive self-image within a group, which is a key part of a reality that gives meaning to our lives.

The principles of social construction are closely related to the idea of "saving face" because both concepts deal with the creation and maintenance of social reality and the importance of recognizing emotions.

Saving Face Is a Social Construct

So many of our breakdowns are connected to losing or saving face. According to social construction, people are constantly "performing" their roles to fit into the shared reality. "Saving face" is a critical part of this performance. It's the act of managing social interactions to avoid embarrassment or loss of status, thereby protecting the socially constructed "self" within the group's reality. If someone loses face, it disrupts the shared understanding of their role and standing.

Shared Meaning: Social construction posits that reality is not objective but rather built through shared language, interactions, and understandings. Similarly, a person's "face" is not an inherent trait but is socially constructed by the respect and dignity they are granted by others. The group agrees on what is considered a "good" or "dignified" status, and people's actions are judged against this socially created standard.

Group Norms: The specific ways people "save face" are themselves social constructs. For example, a collectivist culture might prioritize saving the group's face over an individual's, while an individualistic culture might focus more on personal dignity. These different approaches are based on the socially constructed norms and values of each culture. Managers also must save face with their bosses who expect certain levels of productivity and so forth up the ladder.

In short, "face" is a tangible example of a social construct—it's a concept that has real-world power and influence, even though it's not a physical object. The act of "saving face" is a practical, everyday application of the principles of social construction, as people actively work to uphold and protect the shared social reality in which their self-worth is embedded.

Individuals who work in advisory or leadership positions would be well served to pay attention to saving face for those with power and those without power. No one reacts well to having their identity stripped from them. Only denial can ensue. As a professional seeking to create spaces for open communication, it is most important to be sensitive to saving the face of anyone with whom we interact.

Self-awareness is important here. For so often we are the only ones who know our face is threatened. The point of studying these concepts is to realize that it is far less important to save face than it is to understand the people and situations that we propose to help. I have experienced that letting go of face saving and quickly acknowledging an error or lack of knowledge improves relationships and communication.

Understanding Social Construction and Sensemaking in Risk Assessment

"The perception of risk is a social process" (Douglas & Wildavsky, 1982: 6). Risk is cultural, leading to accountability as a cultural perception as well. In an individualistic culture accountability means individual blame.

Risk is multi-faceted, it's constantly evolving as we encounter new situations, interact with others, and learn from our experiences. This active nature of risk perception highlights the limitations of written rules and procedures, which may not capture the nuances of how people make sense of risk in real-time operational contexts. Therefore, the concept of safety is continuously constructed through the ongoing relationships, interactions, and decisions of individuals within an organization. The gap between compliant design and effective performance arises because regulations and procedures can never fully anticipate the complexities and ambiguities of real-world work environments. A more effective approach to safety involves recognizing its socially constructed nature, engaging in open dialogue, embracing diverse perspectives, and continuously refining our understanding of potential hazards in light of evolving social contexts and practical experiences.

Mann Gulch Wildfire: Sensemaking Gone Wrong

This Mann Gulch wildfire is a stark example of a complete failure of individualism. The firefighters, a cohesive and functional team, devolved into a collection of individuals when their sensemaking and organizational structure collapsed. The crew could not adapt to the unexpected threat, losing their ability to function as a unit.

Weick's investigation found that the tragedy was not caused by a single human error but by a complete breakdown in sensemaking—the process by which people create a shared understanding of their situation.

The Mann Gulch Fire smokejumpers, despite their previous experience with danger, faced an unprecedented situation. Confusing orders and limited communication, coupled with rapidly deteriorating conditions and a lack of strong interpersonal bonds, led to a "every man for himself" scenario.

In a matter of minutes, the situation became so dire that assessment and planning were nearly impossible. Panic set in after they were ordered to drop their tools and then saw their leader light a fire, directing them to enter the burned area. He knew that lying inside a burned area would cause the fire to jump over them. Disoriented and confused, they fled the lurching fire front, facing 30-foot flames advancing at an estimated 660 feet per minute. Sadly, the leader, whose instructions were not understood, was the sole survivor. How or why did this happen? When the fire suddenly changed direction, the firefighters' entire framework of understanding their environment collapsed. Their standard operating procedures became useless, and their leader's command to abandon their tools and join him in an area he had burned instead of fleeing to a safe zone was rejected because it didn't make sense to them. The leader was the only one who knew about the fire jumping over burned areas. He had not shared it with the crew so he lost his authority because he could no longer communicate his orders in a way that made sense to the crew. This is why sensemaking must be a constant exercise. It is the only way to create a shared understanding of risk and prepare for the unpredictable (Hensler, 2024). These concepts of social construction and sensemaking are not merely academic. This high-stakes story forces us to ask a profound question that moves beyond technical failure and into the social anatomy of risk. The failure at Mann Gulch highlights that the most significant risks to life in the workplace are often social and unseen.

The true threat that prevents open communication and adaptive action is not solely the physical hazard, but the fear of social interaction itself. Human interaction has always been and remains the most frequent and riskiest activity in the workplace, where participants risk being slightly embarrassed or deeply humiliated.

This inherent, universal fear—the risk of losing dignity or status—is the core reason people keep their opinions to themselves and fail to challenge authority or confusing directions. Before we can expect people to exhibit psychological safety, we must first explore the pervasive human need to protect their standing within the group: the critical role of "face."

Socially Constructing Safety from an Individual to a Collective Social Outcome

The assertion that risk assessment is a social activity highlights that it is not solely an objective, technical process conducted in isolation. Instead, it is significantly shaped by how individuals interact, communicate, share viewpoints, and collectively interpret situations within a social environment. For executives, understanding this paradigm is critical because customary safety approaches, often rooted in an outdated emphasis on individual responsibility and technical compliance, inherently hinder effective risk assessment and have failed to significantly reduce SIFs.

Adhering to safety approaches that reject or remain ignorant of social construction and the collective interpretation of events debilitates the organization's safety management systems. The intricate web of social, economic, cultural, and organizational factors that shape both risk exposure and behavior cannot be ignored. This process is a continuous and universal phenomenon, constantly shaping our understanding of reality in all human interactions and organizational dynamics. To not learn how to influence these ongoing processes would be akin to running a nuclear power plant without understanding radiation: a potentially catastrophic oversight.

The concept of safety culture is not a static entity but a dynamic social construct, actively shaped through workplace interactions and the collective sensemaking of risk. Understanding how this culture develops is key to risk management.

Many researchers who study disasters emphasize that to decrease SIFs it is imperative to change our attention from the individual to social dynamics within teams and organizations (Flin, 1996; Hopkins, 2005, 2008; Weick, 2024; Rochlin, 1999; Weick, 1988; Leveson, 2004). They collectively challenge the pervasive belief that accidents are primarily caused by individual negligence or carelessness. This involves recognizing how power dynamics, group norms, and individual identities influence safety behaviors. For instance, a hierarchical structure where workers feel afraid to speak up to their superiors can create a breeding ground for accidents. Similarly, a culture where risk-taking is valued and rewarded can lead to unsafe practices becoming normalized.

Social construction is the process by which groups create reality. While exploring social construction as a tool to better understand risk, RCS will serve as the leadership framework for intentionally constructing a reality based on collective safety, trust, and shared meaning. Such a reality aims to employ every worker in the process of identifying and mitigating risks.

These perspectives challenge the common belief that accidents are primarily due to individual negligence. They advocate a social constructionist approach, moving away from conventional, individualistic safety methods. They emphasize that safety is deeply intertwined with social dynamics, organizational culture, and communication patterns, rather than being solely a product of individual actions or technical systems. These scholars collectively argue against the "human error" paradigm, asserting that accidents often result from systemic failures, not isolated individual negligence. Therefore, a comprehensive understanding requires examining the complex interplay of social, organizational, and individual factors (Shufutinsky et al., 2017).

As Rochlin (1999), a pioneering researcher in social construction and high-hazard industries, noted in his groundbreaking paper about safe operations in high hazard industries:

> The maintenance of safe operation … is an interactive, dynamic and communicative act, hence it is particularly vulnerable to disruption or distortion by well-meant but imperfectly informed interventions aimed at eliminating or reducing human error that do not take into account the importance of the processes by which the construction of safe operation is created and maintained.
>
> (Rochlin, 1999)

How can focusing on reducing human error and removing risk become pitfalls? It might seem strange to those who only rely on rules and objective measures. But social construction research shows that safety is often a result of things that are hard to measure, like trust, shared values, and even rituals. When systems are designed to eliminate human input to avoid errors, the humans in charge become passive observers. In fact, Rochlin observed that in nuclear operations rooms, operators became so reliant on their equipment that they lost connection with the holistic operation. This phenomenon is analogous to modern distractions like texting while driving, where focusing on technology can lead to a dangerous detachment from the immediate environment and its evolving risks.

Ronah Flin (1996) specifically highlights communication, teamwork, and leadership as essential elements in constructing and maintaining safety culture, emphasizing the significance of social dynamics within teams and organizations. Flin's "Safety at the Sharp End" emphasizes that robust safety cultures thrive on open and honest communication, enabling the proactive identification and mitigation of risks. Effective teamwork, built on trust and mutual support, cultivates a sense of shared responsibility where every individual can ask questions about safety concerns.

Andrew Hopkins (2005) applied social construction principles to analyze how organizational culture, values, and power dynamics shape risk perception and safety practices. He demonstrated these concepts in his analysis of the BP Texas City refinery disaster how management's values, prioritizing cost-cutting and production, molded employee beliefs about acceptable risk, ultimately contributing to the tragedy. Hopkins argued that operator errors, while present, had deep cultural roots in organizational structure and power imbalances, reinforcing that preventing disasters demands more than technical fixes. His work underscores that safety culture, the existing state of safety, is a product of social construction. The daily peer-to-peer conversations within an organization, significantly influenced by leadership's words and actions, constantly shape safety outcomes.

In sum, these researchers advocate for a holistic safety approach that recognizes the interconnectedness of individual actions, social relationships, and organizational structures, valuing worker expertise, fostering shared responsibility, and acknowledging that safety is actively shaped and maintained through daily interactions and collective beliefs. Table 6.1 summarizes the shift these authors advocate.

Table 6.1 Present and aspirational future of OHS

Present state and mind set	*Shift to aspirational socially constructed and sensemaking view of safety*
Core belief: Safety is objective, static, and predictable; accidents have linear causes	Core belief: Safety is a dynamic, socially constructed achievement; risk is subjective and constantly evolving
Primary focus: Equipment, machinery, physical hazards	Primary focus: Human interactions, social dynamics, communication patterns, lived experience
Checklists, procedures, quantifiable metrics	Shared understandings, collective meaning-making and decision making
Individual responsibility/Human error	Understanding context and conditions
Operating model: Hierarchical control	Operating model: Respectful, collaborative, and participatory
Reactive approach (focus on past events, firefighting)	Proactive and adaptive (anticipating future threats, continuous refinement)
Neglect of emotions and psychological factors	Includes lived experience, motivations, emotions, and interactions in analyzing causes
Consequences to avoid: limited scope (misses social/contextual factors)	Potential benefits: Understand worker perspective of risk
Cause and effect oversimplifies complex situations	Adaptive safety practices, not static
Reduces information flow (stifles worker input)	A common understanding of risk within work groups lessens fear of speaking up
Short-term solutions and recurring incidents	Potential reduction of SIFs due to collective understanding of risk and how to respond
Incident investigations aim to construct a coherent and widely accepted narrative of "what happened" that simplifies complex events into a linear chain of cause and effect	Improve understanding of incidents through inclusion of psychological and social context in investigations to improve the effectiveness of corrective actions

Risk of Social Interaction and Psychological Safety

The act of human interaction has been socially constructed into a high-risk activity. People naturally fear being embarrassed, appearing incompetent, or suffering negative repercussions for what they say or do. To avoid embarrassment or ostracism people spend a lot of time saving face for themselves and others (Baumeister & Leary, 1995; Eagleman, 2011; Vladas etal., 2015; Eisenberger et al., 2003).

The universal human need to "save face" is the primary social barrier to creating psychological safety. Understanding this dynamic is the key to unlocking genuine open communication. This inherent fear is an obstacle for situational awareness when it leads to individuals remaining silent to avoid these interpersonal risks.

Managers are urged to act in ways to create psychological safety. Yet, often, the strongest pressure to remain silent comes from peers, whose opinions and perceptions matter deeply. This dynamic is why a team is only as safe as its least safe member. When individuals perceive sharing ideas or asking questions as too risky, the creation of shared meaning is nearly impossible.

Psychological Safety

Ironically, even "Psychological Safety"—a term meant to describe a group dynamic (Edmondson, 2023) —is being individualized. When managers ask, "Why aren't my employees more courageous *in speaking up?"* they are personalizing the act as a personal responsibility. Consequently, many workplaces struggle with the fact that near misses and errors, even injuries remain hidden for fear of repercussions. Paradoxically, there are stacks of books, training materials, and speakers making it an individual responsibility for managers to create psychological safety.

Psychological safety is a social construction because it is a shared belief among a group that the environment is safe for interpersonal risk-taking (Edmondson, 2018). It is not an innate, individual trait but an emergent property of a group's dynamics, created and maintained through social interaction. It is visible in classrooms or work teams where performance is notably above average. It's a powerful concept because it's a climate that supports learning, creativity, trust, and open communication. The perception of psychological safety is continually reconstructed continuously by social dynamics, including the unwritten rules, power structures, and accepted norms within any group (Bandura, 2001; Thibaut & Kelley, 1959; Edmondson, 2023).

Psychological safety does not mean a lack of conflict or accountability (Coter, 2025). Instead, it supports the harmony of differences, allowing for open discussion and debate. It enables teams to function at maximum capacity by breaking down silos and personal walls that hinder communication and collaboration.

Psychological Safety: A Multifaceted Responsibility

While efforts to create psychological safety often focus on managers, this is a simplistic, individualistic approach that ignores the systemic nature of the problem. Managers are also susceptible to fear and social pressures, and true psychological safety requires a collective effort involving robust policies, peer support, and institutional reinforcement.

Managers may have more formal authority, but they also face greater scrutiny, higher stakes, and a narrower circle of trust. Research shows fear of failure, public embarrassment, or board/market reactions can lead leaders to suppress bad news, avoid confrontation, or resist uncomfortable changes (Kagan & Lahey, 2016,

2020). This fear may be what drives bureaucracy in safety. Layers of complexity are built around safety driven by fear of civil litigation, another example of how power silences all levels, which is the opposite of the psychological safety required for effective risk management.

Creating psychological safety and ensuring adherence to protocols is a complex, systemic challenge, not solely the responsibility of managers. This simplistic view exemplifies the myth of individualism. Both managers and employees naturally follow "paths of least resistance," influenced by fear, incentives, and cultural norms. Overcoming this requires more than individual intentions; it demands robust, information-driven policies, strong peer support, and a shared sense of responsibility.

True psychological safety is not a top-down mandate but a systemic outcome that emerges from the daily actions of everyone involved. While a group dynamic, it is also self-induced. It emerges when an individual's personal criteria for feeling secure enough to be vulnerable are met (Goffman, 1959; Edmondson, 2018, 2018; Bandura, 2001).

Individuals do have unique lenses that influence their initial responses and their personal comfort with vulnerability. However, individual perception is rarely the dominant or final determinant of safety behavior and outcomes in an organizational context. What an individual ultimately does in a safety-critical moment is most often a reflection of the unwritten rules, perceived power structures, and shared understandings that have been socially constructed within their team or organization, even if it contradicts their personal inclination. True psychological safety, therefore, isn't just about an individual's internal criteria but about the social environment that either enables or stifles psychological safety for everyone (Khan, 1990).

Conclusion

This chapter has demonstrated that safety is not an objective, static truth but a social construction. The tragic case of the Mann Gulch wildfire serves as a powerful reminder that when a group's sensemaking collapses, a team of competent individuals can devolve into a collection of isolated actors, with fatal results. This illustrates that our most significant risks are often social and unseen, residing in the very human dynamics that define our relationships. The experience also demonstrates that sensemaking must be practiced all the time, because if you wait until there's an emergency, it will be too late.

We have explored how a singular focus on the individual and technical solutions overlooks these invisible threats. The human needs to "save face", for instance, is a primary social barrier to open communication and a key reason life-saving knowledge can be lost.

Ultimately, safety is a collective achievement. Managing risk in a resilient organization means actively engaging in continuous sensemaking and intentionally creating a culture where psychological safety and trust allow people to speak up without fear. The chapters that follow will build on this foundation by providing the tools and frameworks to make this social achievement a reality.

Executive Takeaways

1 Deconstruct a near-Mmss using a Social Construction lens. Take a recent incident report and use the Mike the Electrician story as a template. Ask the leadership team: What were the power dynamics at play? What unwritten rules or production pressures influenced the decisions made? How was "safety" being defined in that moment by the people involved?
2 Ask managers and supervisors: "Where in our processes is it most difficult for an employee to ask a question or stop a job without feeling embarrassed or disrespectful?" Use the Opticore Tech model of reframing Stop Work Authority as "I've got a question" as a potential solution (Chapter 11).
3 Facilitate a strategic dialogue with the "Present vs. Aspirational Future" (Table 6.1). Use in a leadership offsite to map the organization's current state and build consensus on the aspirational future state.
4 Introduce "Inclusion Precedes Accountability." Make RCS Belief #2 a personal mantra, recognizing that you cannot hold people accountable in a system where they do not feel seen, heard, and included in the conversation about risk.
5 Include sensemaking exercises in repeated routines. These could be *what if* exercises or teambuilding exercises where teams solve potential scenarios.

References

Bandura, A. (2001). Social cognitive theory of mass communication. *Media Psychology*,3(3): 265–299. https://doi.org/10.1207/S1532785XMEP0303_03

Baumeister, R. F., & Leary, M. R. (1995). The need to belong: Desire for interpersonal attachments as a fundamental human motivation. *Psychological Bulletin*, 117(3): 497–529. https://doi.org/10.1037/0033-2909.117.3.497

Chikudate, N. (2009). If human errors are assumed as crimes in a safety culture: A lifeworld analysis of a rail crash. *Human Relations*, 62: 1267–1287.

Collin, R. W. (1992). Book Reviews: Social theories of risk. *Journal of Urban Affairs*, 16(3): 285–294.

Coter, C. (2025). *How to Build Psychological Safety in the Workplace*. Harvard business school online.

Douglas, M. (1992). *Risk and Blame*. Routledge.

Douglas, M., & Wildvasky, A. (1982). *Risk and Culture*. University of California.

Eagleman, D. I.(2011).*The Secret Lives of the Brain*. Vintage Books.

Edmondson, A. C. (2018). *The Fearless Organization: Creating Psychological Safety in the Workplace for Learning, Innovation, and Growth*. Wiley.

Edmondson, A. C., & Bransby, D. P. (2023). Psychological safety comes of age: Observed themes in an established literature. *Annual Review of Organizational Psychology and Organizational Behavior*, 10(1): 55–78. https://doi.org/10.1146/annurev-orgpsych-120920-055217

Eisenberger, N.I., Lieberman, M. D., & William, K. D. (2003). Does rejection hurt? An MRI study of social exclusion. *Science*, 302:290–292.

Flin, R. (1996). *Sitting in the Hot Seat: Leaders and Teams for Critical Incident Management*. Wiley.

Goffman, E. (1959). *The Presentation of Self in Everyday Life*. Anchor Books.

Hensler, B. (2024). Mann Gulch revisited: A world turned upside-down. *Fire Rescue.* www.firerescue1.com/wildfire/articles/mann-gulch-revisited-a-world-turned-upside-down-PHb5vbr5zKfQWVpO/

Hopkins, A. (2005). *Safety, Culture and Risk.* CCH

Hopkins, A. (2008). *Failure to Learn: The BP Texas City Refinery Disaster*. CCH Australia LMT.

Kagan, R., & Lahey, L. L. (2016). An everyone culture. *Harvard Business Review Press.*

Khan, W. A. (1990). Psychological conditions of personal engagement and disengagement at work. *The Academy of Management Journal*, 33(4): 692–724.

Leveson, N. (2004). A new accident model for engineering safer systems. *Safety Science*, 42(4): 237–270. https://doi.org/10.1016/S0925-7535(03)00047-X

Mead, G. H. (Downloaded 2025). G.H. Mead and self identity. *ChromeIAS.* https://chromeias.com/g-h-mead-self-and-identity/#:~:text=Mead%20(%20GEORGE%20HERBERT%20MEAD%20)%20argues,is%20able%20to%20look%20back%20upon%20himself

Nichols, T. (1975). *The Sociology of Accidents and the Social Production of Industrial Injuries: People and Work.* Holmes McDougall Edinburgh.

Nichols, T. (1999). Death and Injury at Work: A Sociological Approach. In: Daykin, N., & Doyal, L. (Eds). *Health and Work.* Palgrave.

Rochlin, G. I. (1999). Safe operation as a social construct. *Ergonomics*, 42(11): 1549–1560.

Schein, E. H. (1992). *Organizational Culture and Leadership*. Jossey-Bass.

Shufutinsky, A., Vizcarrondo, M., & Cox,R. (2017). The collapse of sensemaking in injury root cause investigations resulting in ineffective injury prevention decision-making: A retrospective case study. *Environmental Science*. www.semanticscholar.org/paper/The-Collapse-of-Sensemaking-in-Injury-Root-Cause-in-Shufutinsky/18ffd88b2e8a0671912b3ca5aaa0d6f4ea4ca4a6

Taylor, D. H. (1981). The hermeneutics of accidents and safety. *Ergonomics*, 24: 487–495.

Tetteh, C. (2025). *Organizational Learning through Storytelling*. Department of Engineering, University of Alabama Birmingham. ASEM 620: Capstone Project Part 2.

Thibaut, J. W., & Kelley, H. H. (1959). *The Social Psychology of Groups*. Wiley.

Tierney, K. J. (1999). Toward a critical sociology of risk. *Sociological Forum*, 14: 215–242.

Trevino, J. A. (Ed). (2003). *Goffman's Legacy*. Rowan & Littlefield.

Turner, N., & Gray,G.C. (2009).Socially constructing safety. *Human Relations*, 62(9): 1259–1266.

Vladas, G., Haselton, M. G., & Ackerman, J. M. (2015). Evolution and close relationships. In: Mikulincer, M., & Shaver, P. R. (Eds). *Vol. 3: Interpersonal Relations.*

Weick, K. (1988). Enacted sensemaking in crisis situations. http://wendynorris.com/wp-content/uploads/2018/08/Weick-1988-Enacted-Sensemaking-in-Crisis-Situations.pdf

Weick, K. (2024). Mann gulch revisited: Improvisation as a surface of apprehension. *Academy of Management*. https://doi.org/10.5465/amd.2024.0006

Zenger, J., & Folkman, J. (2017). Why Do So Many Managers Avoid Giving Praise? *Harvard Business Review*. https://hbr.org/2017/05/why-do-so-many-managers-avoid-giving-praise

7 Using Language as a Strategic Tool

The language of leadership is not a "soft skill" but a strategic tool that fundamentally shapes an organization's reality, culture, and safety outcomes (Mallinson, 2025). Language is the primary tool we use to construct and communicate meaning (Stein & Ridderstråle, 2003). We create shared meaning through sensemaking in conversations and dialogue. Cultural and social context influences how we interpret meaning, and those with more power often have a greater influence on shaping dominant meanings. Language does this by making concepts visible and defining what exists, what is true, and what is allowed. Nothing is visible until it has a name. Consider the mid-20th century emergence of the term "depression"; the new concept fueled a multibillion-dollar antidepressant industry by 2024.

This chapter provides the evidence, theory, and practical steps to demonstrate how intentional linguistic choices are a core competency for executive leadership. Language is not a conveyer for your strategy; it is your strategy. If you change the words, you change the world.

Introduction

Our insights begin with Professor Lera Boroditsky's research, which asks: "Does the language we speak shape the way we think?" According to her, the Holy Roman Emperor Charlemagne said, "To have a second language is to have a second soul," implying that each language provides the brain with another way of seeing the world. Her research shows that words shape human knowledge and behavior to solve problems (Boroditsky & Gaby, 2010). For instance, the Russian language differentiates between light and dark blue, enabling Russian speakers to perceive and categorize shades of blue faster than those whose language has only one word for it (Boroditsky, 2018, 2020). This isn't because their eyes see differently, but because their language teaches their brains to categorize these distinctions more readily.

By demonstrating that language shapes how we perceive and categorize the world, Boroditsky's research implicitly challenges the idea of purely objective definition of risk. If our very perception of risks and hazards is influenced by the linguistic frameworks we operate within, then an individual's action is not entirely self-directed but is shaped by the linguistic environment.

DOI: 10.4324/9781003668770-7

In short, words actively create reality. Language makes concepts visible and defines what exists, is true, and is allowed (Luhmann, 2003). Once a concept has a name, it becomes tangible, influencing collective focus and investment.

This research could have implications for safety education and training. For example, if a student is focused on procedural adherence, it limits the scope of their risk perception much like language with only one word for blue. Could this create a cognitive bias where individuals are only trained to look for problems that fit within the framework they were taught? This might limit their ability to perceive unforeseen risks or problems that fall outside of the established procedure, as they don't have the linguistic or conceptual tools to categorize them (Flin, 2017).

The Cost of Careless Language: Case Study in Organizational Failure

Careless leadership language actively constructs a reality where risk is normalized, often with tragic consequences.

People are affected by the mindset that language embodies. A pervasive individualistic mindset can subtly yet dangerously erode collective safety by supporting a self-centered approach to work. It prioritizes one's own immediate tasks or limited departmental view over the well-being of the entire workforce. There is little consideration of the ripple effects across the whole system or the collective human cost. A visceral example:

> On November 9, 2003 an article appeared in the *Los Angeles Times* entitled, "Disney Ride Upkeep Assailed." In 1997 Disneyland moved to what is known as "reliability-centered maintenance" to save money. Three employees said they heard the park president say, "We have to ride these rides to failure to save money." Then, for the first time in the park's history two fatalities and ten injuries occurred because of equipment failure between 1998 and 2003.
>
> Outside experts attest that Disneyland is a safety leader among amusement parks, but according to the article, many of the workers interviewed say that the move "gutted worker morale and employee's sense of ownership of the rides."
>
> Officials said that the complaints came from disgruntled workers, but one supervisor who worked at Disney from opening day to 1997 when he retired said, "When Disneyland opened, safety was the No. 1 thing. Now they say that today too. But I think over time, profit became more important".
>
> (Anton & Yoshino, 2003a)

As a former mechanic recalled, when a consultant questioned why lap bars on a ride that had never failed were checked daily, the supervisor reacted with disbelief: "The reason they don't fail," he said, "is because we check them every night" (Anton & Yoshino, 2003b; Dowd, 2022). President Pressler had one outcome in mind, but his words had unintended consequences because of the way they were interpreted by the workforce.

This case isn't an isolated incident. Chapter 5 detailed the mining disaster at Massey Energy's Upper Big Branch (UBB) mine, the CEO encouraged supervisors to focus on "running coal" rather than complying with safety standards, leading to the deaths of 29 miners. In each instance, individualism played a role by prioritizing profit for a small group at the expense of the many.

The Linguistic Divide in Management

An analysis of language used by different managerial groups in this study reveals a profound linguistic divide that constructs two separate organizational realities. The findings suggest that for middle and frontline management, experience shapes the language they use, while for executives, the way they think shapes their language. The language for the study was collected during interviews for a safety culture assessment after a substantial personnel layoff followed by a spike in accidents.

The narrative responses were analyzed for the words most frequently used and emotional content. Table 7.1 captures the most frequent terms and concerns. They were asked an open question, "What needs to be addressed to improve safety and health at this company?"

Analysis of Middle Management and Executive Language

Table 7.1 reveals a collision of two worlds. Executives are playing a game of "Assets" and "Mitigation," while the frontline is fighting a war of "Burnout" and "Stress." They aren't just using different words; they are solving different problems. Much like Boroditsky's subjects who lacked specific color terms, the Executives' reliance on abstract nouns like 'Asset Health' renders the 'human-centric' hazards—such as burnout—cognitively invisible to the strategic layer of the organization.

The table clearly demonstrates the concrete or lived experience of supervisors and managers versus the abstract and distanced experience of the executives. As discussed, our understanding of concepts like "safety" and "risk" is not inherent but is shaped by social interactions, shared language, and cultural context. This is evident in the distinct language used by different managerial groups. The language of middle managers and executives, rather than merely describing reality, actively constructs their respective organizational realities and priorities.

Executives use high-level, abstract language (e.g., "risk appetite," "asset health"), while middle managers use tangible, employee-level language (e.g., "well-being checks," "hands-on training"). The middle management solutions emphasize people-centric solutions while the executives emphasize systemic long-term strategies and organizational governance.

Executives spoke the language of "assets" and "governance," while managers were talking about "sweat" and "stress." When an executive asks for "risk mitigation" (a spreadsheet task), the manager hears "more paperwork," not "safer workers." Executive use of high-level abstract concepts often fails to provide a

Table 7.1 Comparison of safety priorities between middle management and executive level

Middle management and supervisors priorities	*Executives priorities*
Creating a safer and stress-free working environment	Safety culture
Addressing frequent rotational schedules	Risk management
Work–life balance	Risk appetite
Better leadership and focus on people	Risk mitigation
Regular site visits and compliance reviews	Safety standards
Improved medical coverage for employees and their families	Incident management
Continuous timely delivery of health and safety e-learning courses	Root cause analysis
Enhanced healthcare benefits	Desire for a proactive rather than reactive approach to risk
Safety support systems needs	Concern about unclear accountability for safety
Easy access to management systems in the field	Need for better incident investigation processes
First line resourcing	Asset management and infrastructure
Regular interactions with leaders and individual well-being checks	Asset health
Work from office flexibility	Infrastructure management
Hands-on training improvement	Network infrastructure
Pressure to maintain goals and sales validation compliance	Physical assets
Distinction between health, safety, and well-being	Aging infrastructure
Property maintenance and resolution of issues	Asset health uplift
More 1-on-1 training instead of e-learning	Concern about maintaining and upgrading aging infrastructure
Availability of a health, safety, and well-being hotline	Recognition of our focus on asset management rather than technology

clear, collaborative roadmap for the middle managers and employees to achieve executive priorities.

For instance, executives express a desire for a "proactive rather than reactive approach to risk." However, their language is heavily focused on reactive measures such as "incident management," "root cause analysis," and "risk mitigation," which describe post-incident actions. In contrast, middle managers use more proactive terms like "hands-on training improvement," "regular interactions," and "well-being checks," which represent behaviors that prevent incidents.

Similarly, executives voice "concern about unclear accountability for safety," yet their own language contributes to this problem. They employ broad, systemic terms like "safety culture" and "risk management" without linking them to specific examples, roles, or responsibilities, leading to vague and diffused accountability. Middle managers, conversely, discuss direct actions and responsibilities, including "site visits," "compliance reviews," and "1-on-1 training." The executive's language lacks the embodiment or emotional context necessary for implementing these specific tasks. They don't acknowledge individual stress, constructing a reality where safety is a strategic asset to be managed, not a personal experience to be supported.

Ultimately, the executive's strategic language, while important, lacks the specific, actionable terms required to enable collaboration and collective responsibility. Their top-down, command-and-control vocabulary is detached from the operational reality.

Impact on Collaboration and Co-creation

The executive's language is not just a different dialect; it's a barrier to communicating a shared vision to non-executives. Their language creates a perception that safety is a theoretical, top-down mandate rather than a collaborative effort. It suggests a separation between "us" (the executives who set the strategy) and "them" (the managers and employees who execute the tasks). The two groups are talking past each other because their maps don't match. The executives describe a system of concepts while the managers describe a system of people. They are not just differing on priorities, they are inhabiting different realities.

The different lexicons prevent a shared understanding of what "proactive safety" and "accountability" mean in practice. For an executive, it may mean a new risk matrix. For a middle manager, it might mean having the authority to pull a worker off a dangerous task without fear of reprisal. So much emphasis is placed on accountability, but there is very little agreement on how it emerges. Relationship-Centered Safety (RCS) Belief #2: "Inclusion precedes accountability" points to the significant implications of this divide. The prevention of SIf's must be a collaborative human centric effort. Meaning that people do not tend to take ownership of responsibilities imposed on them where they had no voice.

An Occupational Health and Safety Advisor Shares How Words Make a Person Feel Included:

> When I did site inductions at my old company, I would ask people what their roles were and would get the response "just a labourer" a lot. My response would always be, 'you are not just a labourer, you are an integral part of keeping the site running.' You could immediately see the appreciation on their face as at previous sites they had mostly been treated as expandable and beneath others.
>
> (Richard Knowles, former DuPont manager interview)

People know when they are seen as a person of value. They also know when they aren't even if management says the right words. This was expressed in a supervisor's survey comment:

> Executives talk about well-being and mental health but the reality of working here is completely different. For some time now, the number of staff members who use the words burnout, exhaustion, stressed out regularly, is something I have not previously experienced. Yet, I don't see any response.
>
> (Supervisor interview)

Executives tend to hold a broad, abstract view of the organization, while frontline supervisors and their managers possess detailed, concrete knowledge of daily processes and hazards. This came through in the survey comments. People are expressing the need for more staff, replacing equipment, or more breaks to reduce stress. Yet, none of the upper-level managers brought this up in their interviews. When these differing perspectives aren't carefully sought out and discussed, top-level decisions may lack practical relevance, and ground-level issues may not be understood or escalated appropriately.

A Human Dimension of Risk: Embodied Cognition

We have described the linguistic divide between executives and middle management and some of the barriers it creates in safety communication. This section will talk about how embodied cognition drives this divide and why frontline language is so critical. Embodied cognition argues that the body's physical experiences provide the "meaning" for these abstract ideas. Thus embodied meaning is the theory that meaning and cognition are inherently rooted in our bodily experiences and physical interactions with the environment, challenging the idea that mind and body are separate.

Embodied cognition is strongly related to emotions, which are vital to human survival (Massey, 2002). Damasio's (1994) experiments revealed that individuals with brain damage that severed the connection between their emotions and the prefrontal cortex could not make even simple decisions. This makes emotions essential for navigating life's complexities and in daily decision-making. Emotions also affect our comprehension. Abstract concepts aren't likely to generate bodily reactions, but the mention of a lemon can make our mouth pucker. This is because the body remembers the feeling associated with the word. An emotional concept can also involve a physical reenactment of a bodily state associated with that emotion. This is where empathy resides.

These concepts are not academic. They are foundational to two aspects of social competence especially relevant for senior managerial influence on safety: social perception and persuasion. Gardner and Stough (2002) emphasize the ability to understand the emotions of others as being among the strongest predictors of senior managers' effectiveness and this has also been highlighted for the influence on safety (Hopkins, 2012). A manager's communication and active involvement are strengthened by the ability to read people. Therefore, the ability to understand the emotions of others (social perception) is one of the strongest predictors of a senior manager's effectiveness and their influence on safety.

Being able to read people's body language and expressions is a key leadership skill because embodied sensemaking is the primary way that people use to identify risk in the field. A 2024 study explores embodied sensemaking, examining how the physical senses of frontline workers—what they see, hear, and feel—serve as inputs for detecting weak signals and initiating the sensemaking process long before formal analysis begins. This work argues for understanding sensemaking as an embodied, socio-material process, not just a cognitive or social one (Bower et al., 2018).

Embodied sensemaking is the primary method by which frontline workers identify risks, making the ability to interpret people's body language and expressions a vital leadership competency. A 2024 study highlighted this by showing how the physical senses—what workers see, hear, and feel—are critical inputs for detecting "weak signals," initiating the sensemaking process well before any formal analysis starts. This research frames sensemaking not merely as a thinking or social activity but as a physical interaction with the environment (tools, equipment, weather) (Bower et al., 2018).

Embodied Cognition as a Sixth Sense

Our personal physical and sensory experiences determine how we understand and communicate risk. Embodied cognition emphasizes that risk perception is context-dependent. For example, someone who has personally felt the intense heat of a near-miss fire or the discomfort of ill-fitting personal protective equipment will discuss these risks very differently from someone who has only read about them. Tapping into this ability is important to effective communication about hazards or why a procedure is important. People are much more likely to understand and adopt ways of doing things that "feel right."

Weick (2001) and Johnson (2017) argue that understanding isn't solely a logical process. Rather, it's deeply intertwined with our interaction with the environment and our bodily and emotional experiences. New Zealand's Scion's research institute applies this framework to investigate decision-making and its potential influence on the future of safe practices. The Scion team observed that skilled workers operating in uncertain environments progressively develop a "sixth sense." Their body learns to unconsciously process incoming information. This enables automatic responses, freeing up attention resources to concentrate on processing unexpected events. The underlying idea is that the body's interactions with its environment contribute to cognition, which is the mental action or process of acquiring knowledge and understanding through the senses (Hooper et al., 2022).

Use the Embodied Language of Your Audience's Lived Experience

When company communications ignore these basic aspects of how humans learn and store information, they speak in technical or neutral language. That risks alienating listeners who feel unheard and devalued. It also perpetuates an ingroup/outgroup status—"only those in the know understand."

Table 7.2 Using the language of the listener's lived experience

Question	*Company answer*	*Critique*	*Constructive language*
We have a chilled ("cold") working environment where people are afraid to raise questions or concerns due to fear of retaliation.	We have a genuine desire to hear from you. The leadership team continues to explore multiple avenues to allow open and honest dialogue with employees, managers, and supervisors.	Missed the opportunity to address the embodied cognition of a "chilled work environment" and the fear of retaliation. Without that connection there is little possibility that the "genuine desire to hear from you" is believed.	We aim for a warm, welcoming place where people can speak freely. We continue to engage with supervisors, managers, and directors to set expectations to create an open, learning environment. If you have any questions or concerns regarding retaliation, please contact your organization's labor relations advisor or human resources to make a report, or call anonymously phone #...

Note: See Appendix B for more examples of the question, initial responses, and why they were likely ineffective.

Employees often use vivid, emotionally charged language to describe the direct impact of workplace conditions on their physical and mental well-being. It is an embodied meaning that arises from the tangible, felt experiences of their daily work lives. Someone using that language is unlikely to understand or connect with abstract, non-emotional responses.

For example, an employee's response to why he lost trust in management was "After I had an accident, they threw me under the bus." Unfortunately, management's response was, "We realize we have to work on a consistent response to incidents. We should have something in six months." The employee didn't have a good thing to say after that conversation. The differences in language demonstrate that managers and employees live in different realities.

A meaningful response would have addressed the underlying emotional concern first with embodied language to connect at the emotional level. Table 7.2 provides an example of a meaningful response.

Case Study: The Rise of Embodied Cognition over Risk Metrics

Tetteh (2025) documents an example in his master's thesis of how lived experience and embodied cognition can be more accurate than technology.

A large underground mining operation prided itself on its state-of-the-art risk management system. Their safety protocols were meticulously crafted, relying on sophisticated models, continuous monitoring, and quantifiable data. Management believed their system provided an objective, rational, and comprehensive understanding of all potential hazards.

One evening, Alex was tasked with installing roof bolts in a new section of the mine. Despite a methanometer reading "all green," his internal alarm went off. He detected a "faint, sweet, cloying smell" and an overall sense that "the air had a different taste." He instinctively knew they were approaching an unmapped methane pocket.

He immediately stopped work and notified his supervisor. The supervisor, adhering strictly to the "green zone" readings and under pressure for production, dismissed Alex's concerns. Yet, Alex held firm, stating, "The system doesn't smell the air or know the stories my grandfather told me about these rocks." The supervisor reluctantly called in engineers and the special gas team. Within minutes, they confirmed Alex's intuition: a significant, previously undetected methane pocket was indeed present. Had Alex pushed through, a catastrophic event would have occurred.

Analysis of Alex's Case

Alex's story directly challenges the notion of risk as a purely technical, quantifiable measure. While management relied on "calculated allowances" and "green zone" models, his perception of risk was deeply visceral, sensory, and experience-based. His "faint, sweet smell" and feeling that "the rock is breathing differently" were forms of lived experience, local knowledge, and tacit understanding gained over decades and through generational stories—knowledge entirely absent from the sophisticated risk models. This highlights the disparity between management's perception of risk, often a calculated "allowance," and employees like Alex who see risk in a more visceral, sensory reality.

The initial dismissal of Alex's concerns by his supervisor further exposes how power dynamics can lead to the suppression of critical frontline awareness of hazards when it conflicts with production targets or management's top-down control over risk definitions.

A New Lexicon: Case Study in Language Transformation with the 3-Cs

The language used in organizations frames issues and sets the level of psychological safety. Using Positive/Empowering Framing ("Your observations are valuable and help us continuously improve our shared safety environment") is key to nurturing the trust required for open reporting and collective learning.

The following case study demonstrates how to apply the lessons learned about embodied Cognition to actively build a relational culture. These seemingly small linguistic changes further the engagement and build the trust necessary for a resilient safety culture, embodying the principle that culture doesn't shift because people read something; it shifts because they experience something.

Background

A nuclear power plant (NPP) executive coach shared the following process he used successfully at an NPP. As you read this case study of the NPP manager who introduced the "3-Cs" (Caring, Collaboration, Co-creation), note that it is a perfect example of a leader using language to construct a collective, relational reality. This is RCS in action, directly countering the abstract, detached language of individualistic management systems.

The nuclear power industry is a high-stakes world. During outages, the pressure to perform—quickly, flawlessly, and safely—is immense. I've seen it many times: the default mode often leans toward command, control, and compliance. It's efficient, yes, but it can also create an environment rooted in fear, not engagement. My goal as a coach is always to help leaders see beyond that default, to tap into the enormous potential that lies in human connections and relationships. I aim to turn leaders into mentors. You want to be curious, present, and affirming.

That's precisely the challenge I faced with an outage manager at a nuclear plant. He was good at his job, technically proficient, and driven by the pressure of these critical events. Crucially, he bought into my ideas from the start and genuinely wanted to collaborate. He recognized that while their existing methods got the job done, there was a deeper level of performance and engagement to unlock. He was eager to learn, and that willingness to embrace a new approach made all the difference. My strategy revolved around what I call the 3-Cs: Caring, Collaboration, and Co-creation.

We started small, right in the daily 8 AM Leadership Alignment Meeting. This meeting was a cornerstone of their operations, bringing together 800–900 craftspeople, supervisors, and assistant plant managers from across the country. It was the perfect stage to introduce subtle, yet powerful, changes.

My first suggestion to the manager was simple: "When you respond to someone, say their name." It sounds incredibly basic, doesn't it? But the impact is profound. Saying someone's name pulls them into the present moment. It affirms their presence, connects them, and engenders a sense of belonging. Think about it: when someone is spiraling out of control, the first thing you do is try to ground them by saying their name. This manager, for all his experience, had never considered something so self-evident.

The very next meeting, he put it into practice. "Thank you, Rosa, I really appreciate that feedback." It was a small shift, but his team noticed immediately. People started stopping him in the hallway, genuinely curious. "God, what is your coach doing with you? You're so different!" The early signs were incredibly encouraging.

As the first outage progressed, he began to explicitly affirm the team's effectiveness, not just the result but the process. He'd highlight how seamlessly they handled a handoff, or how well they anticipated another team's needs, or simply how effectively they were working together. This wasn't about empty praise; it was about recognizing specific, collaborative behaviors, and it visibly boosted the team's self-esteem.

After that outage, we delved deeper into the language of the 3-Cs. I specifically coached him on how to introduce collaboration. It was straightforward: after a team member reported a successful interaction, he was to use the word collaborative to describe the action. For example, if someone said, "We had a really good handoff on the third shift last night. We caught all the action items and we got ready for the team," his response would be, "That's great, Jim, that's very collaborative." This directly linked the behavior to the concept, and before long, "collaborative" became part of their everyday language.

I introduced the word care when I was coaching one of the outage managers who was having a lot of issues with her team. Leading an all-male team was difficult, and she found herself in many conflicts. I suggested that she began communicating first about why she cared about the process she was presenting or the approach. I told her she could also find ways to communicate that she cared for each member of her team, their ideas, and who they were. By the end of that outage, her team was working together smoothly and she was praising their teamwork.

Finally, we introduced co-creation in the context of planning for the next shift. Instead of just "planning," they started saying, "We need to co-create this." This particular shift in language had the most powerful impact. Co-creation is about shared purpose and vision, and when a team aligns around a common goal, the collective power they unlock is truly remarkable.

Witnessing this transformation was incredibly rewarding. It showed that even in high-pressure, rigidly structured environments like an NPP, a shift from control and fear to caring, collaboration, and co-creation isn't just possible—it's powerful. It creates a workforce that's not just compliant but genuinely engaged, effective, and deeply connected. The manager's openness to learning and his desire to partner in this cultural evolution were keys to its success.

Analysis of Social Construction in the 3-Cs Case Study

The 3-Cs worked because it wasn't a memo. The consultant didn't ask the manager to "implement a paradigm"; he asked him to say a name. This linguistic analysis is pivotal; it labels and critiques the dominant discourse, making its flaws visible. By introducing the new 3-Cs, the consultant offers an alternative linguistic framework for interaction. This isn't merely about adopting new words; it's about shifting the underlying assumptions and power dynamics embedded in those words. "Caring" replaces hierarchical detachment, "collaboration" replaces top-down directives, and "co-creation" replaces mere obedience.

Simple acts, like using someone's name, are highlighted as powerful tools for "present moment awareness," affirmation, connection, and belonging. The consultant articulates the psychological effect of such seemingly minor linguistic choices, explaining how calling someone by their name or bringing up something you care about can pull individuals from a state of being "out of control" into a state of engagement and recognition.

The client's adoption of this practice and the subsequent positive feedback from his team ("You're so different") provide empirical evidence of language's transformative power at an interpersonal level. This works because it isn't a memo. The consultant didn't ask the manager to "implement a new paradigm"; he asked him to say a name. By replacing the detached language of compliance with the intimate language of caring, they didn't just change the vocabulary—they changed the power dynamic.

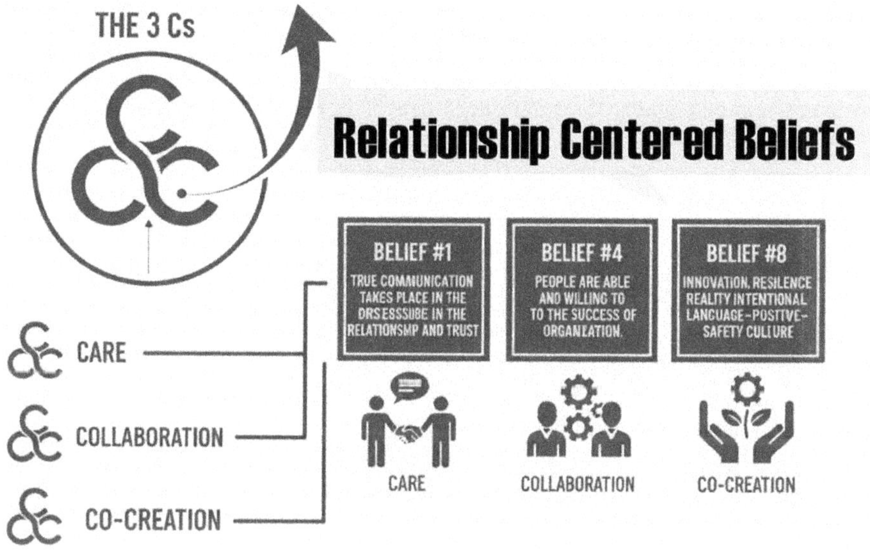

Figure 7.1 The 3-Cs and eight beliefs.

The 3-Cs and RCS Beliefs

Using the 3-Cs to develop team trust, collaboration, and open communication is also an example of how to put the RCS beliefs in action. Each time the words were consciously applied, it triggered a sense of belonging and being valued. The ultimate results from this approach demonstrate Belief #3: Innovation, resilience, inclusion, and accountability are interdependent. Letting someone know you care creates the *inclusion* necessary for resilience.

Communication about safety isn't just about transmitting objective information; it's a much richer and more complex process. It includes our actions, body language, tone of voice, facial expressions, and even shared physical experiences. For example, a leader's calm demeanor during a crisis communicates safety as much as their verbal instructions. The meaning of safety is constantly being refined and reshaped as people interact, learn, and adapt to new situations. This is the social construction of meaning in action.

Language is a powerful tool that can be used positively as demonstrated with the 3-Cs. However, it can also be used to reinforce societal biases and stereotypes. For example, if an older employee expresses concern about working too many hours, they might be labeled as a liability. This can lead to discriminatory hiring practices. Similarly, euphemisms like "man-hours lost" instead of "workers injured" depersonalize the event and diminish the impact on human life. Acknowledging language's creative power and using it intentionally is powerful tool for cultivating a human-centered and safer workplace that recognizes the emotional aspects of human experience. Understanding that your words actively construct reality enables the intentional use of language to cultivate a positive safety culture.

Conclusion

We began this chapter by asking if language shapes reality. The application of evidence—from Boroditsky's research to the tragic "Ride to Failure" at Disneyland—indicates that it does. We have seen that words are not neutral; they are the architectural blueprints of your safety culture.

The data reveals a dangerous Linguistic Divide: executives speak in abstract systems ("asset health"), while the frontline speaks in visceral experiences ("sweat and stress"). As the Alex case study in the mine proved, this gap blinds leadership to the "weak signals" that only human senses can detect—a risk that will only grow as we rely more on AI and remote management.

However, the gap can be bridged. The nuclear plant case study demonstrated that by shifting from the language of control to the 3-Cs, leaders can dismantle the hierarchy of silence. If you change the vocabulary, you change the power dynamic.

In the story of Alex in the mine, a calibrated sensor might read "all green," but a human body can smell the "sweet, cloying" scent of danger. However, when workers are trained to ignore their own sensory inputs, it leads to a chilling deduction regarding the future of technology: AI and remote management represent the ultimate safety risk.

The digital dashboards or AI protocols are blind to the "weak signals" that only physical presence can detect. The AI also speaks the dialect of the Executive—abstract and logical—thereby widening the gap between the plan and the reality of the floor. The "Ride to Failure" metaphor at Disneyland confirms this division.

The solution found in the 3-Cs model proves that the organization is its conversations. The nuclear plant manager who started using people's names and labeling actions as "collaborative" didn't just change his vocabulary; he dismantled the hierarchy of silence.

Ultimately, this analysis forces a redefinition of leadership. If an organization is built on relationships, then "small talk" is not a waste of time—it is "creating the relational network" that creates trust. A leader who refuses to engage in the embodied, messy language of the frontline is not being "professional"; they are actively dismantling the only detection system that can save them from the next catastrophe. By shifting from the language of control to the language of co-creation, leaders cede power to gain the one thing their dashboards cannot give them: an embodied version of the truth (Carrillo, 2019).

Executive Takeaways

This chapter's central message is this: language is not a "soft skill" but a strategic tool that actively shapes your organization's reality, culture, and risk profile. A failure to manage language is a failure to lead, with direct consequences for the bottom line and business continuity.

- Use the "Linguistic Divide" analysis as a diagnostic tool. Review recent all-hands meeting transcripts or emails and assess whether the language used was abstract and systemic or embodied and relational.
- Pilot the "3-Cs" Model. Identify a single, high-leverage meeting (like a daily operations stand-up) and begin practicing the simple linguistic shifts from the nuclear plant case study. Start by using people's names and intentionally labeling positive actions as "collaborative."
- Invest in leadership development that focuses on social and problem-solving skills, specifically training leaders to recognize and use embodied language. Your Leadership Team Leaders must learn to speak the language of the frontline to build trust and gain accurate information. Embodied cognition is the concept that understanding is tied to physical and emotional experience.
- Prioritize leadership development that focuses on social perception—the ability to understand the emotions and intentions of others. Train leaders to recognize and use embodied language to connect with employees on an emotional level, making them feel heard and valued.
- Despite its power to shape thought, language cannot solely drive change. While a deliberate and authentic change in language is a powerful foundation for reshaping the organizational safety mindset, it is merely the critical first step. True and lasting safety improvements require this linguistic shift to be supported by genuine behavioral changes, aligned resource allocation, clear policies, consistent leadership, and robust systems to manifest as real-world results and avoid cynicism.

References

Anton, M., & Yoshino, K. (2003a, November 9). Disneyland's ride upkeep criticized by park workers. Los Angeles Times.

Anton, M., & Yoshino, K. (2003b, November 27). State blames Disneyland ride fatality on crew.

Boroditsky, L. (2018). *7,000 Universes: How the Languages We Speak Shape the Ways We Think*. Doubleday Canada.

Boroditsky, L. (2020). How language shapes thought. *The Long Now Foundation*. www.youtube.com/watch?v=I64RtGofPW8

Boroditsky, L., & Gaby, A. (2010). Remembrances of times east: Absolute spatial representations of time in an Australian aboriginal community. *Psychological Science*, 21(11): 1635–1639.

Bower, M. W., Gylstorff, S., Bargmann Madsen, E., & Mejlgaard, N. (2018). The Fukushima accident and public perceptions about nuclear power around the globe: A challenge and response model. *Environmental Communication.* http://eprints.lse.ac.uk/88103/7/Bauer_Fukushima%20Accident_Accepted.pdf

Carrillo, R. A. (2019). *The Relationship Factor in Safety Leadership*. Routledge.

Damasio, A. (1994). *Descartes' Error: Emotion, Reason, and the Human Brain.* Penguin Books.

Dowd, K. (2022). Death in Disneyland: Remembering the park's most turbulent era. www.sfgate.com/disneyland/article/the-most-turbulent-era-in-disneyland-history-17125238.php

Flin, R. (2017). Enhancing safety performance: Non-technical skills and a modicum of chronic unease. In Bieder, C., & Bourrier, C. (Eds.), *Trapping Safety into Rules: How Far Can We Go?* Springer, 139–153.

Gardner, L., & Stough, C. (2002). Examining the relationship between leadership and emotional intelligence in senior level managers. *Leadership and Organization Development Journal*, 23: 68–78.

Hooper, B., & Parker, R. (2022). Capturing & harnessing embodied cognition problem & opportunity. www.researchgate.net/publication/383060977_Capturing_Harnessing_Embodied_Cognition_Problem_Opportunity#:~:text=Abstract,%2Dthe%2Djob%20learning%20approach.

Hopkins, A. (2012). *Disastrous Decisions*. CCH Australia.

Johnson, C. E. (2017). *Meeting the Ethical Challenges of Leadership: Casting Light or Shadow* (6th ed.). Sage Publications, Inc.

Luhmann, N. (2003). Society's structure and semantics. In Rasch, W. (Ed.), The Making of Meaning: From the Individual to Social Order: Selections from Niklas Luhmann's Works on Semantics and Social Structure. Stanford University Press. 162–234.

Mallinson, K. (2025). *Sociolinguistics*. Britannica. www.britannica.com/science/sociolinguistics

Massey, D. S. (2002). A brief history of human society: The origin and role of emotion in social life. *American Sociological Review*. 67(1): 1–29.

Tetteh, C. (2025). *Organizational Learning through Storytelling*. Department of Engineering, University of Alabama Birmingham. ASEM 620: Capstone Project Part 2.

Weick, K. (2001). Tool retention and fatalities in wildland fire settings: conceptualizing the naturalistic. In Salas, E., & Klein, G. A. (Eds.). *Linking Expertise and Naturalistic Decision Making* (1st ed.). Psychology Press.

8 Social Construction and Sensemaking in Risk Management

Introduction

Safety is a social achievement (Sutcliffe, & Barton, 2009). Building on the insights from social construction and sensemaking established in Chapter 6, this chapter applies them to how we approach risk. While the previous chapters deconstructed the myth of individual responsibility, this chapter links the myth to the concept of human error. If you believe in one you must believe in the other. The chapter, then, moves on to provide the practical framework for reconstructing a more human-centric approach to risk management.

Principles of Social Construction

The six principles of social construction (Table 8.1) are a diagnostic tool to conduct risk assessments and audits. The principles have proven to increase the understanding of risk and aid in the application of these skills during risk assessments, safety audits, conversations, shift turnovers, and operational and safety meetings.

Integrating these principles into leadership practices can create a more human-centered safety approach with the goal of effective risk management, improved safety performance, and a more engaged and empowered workforce.

The Influence of Society and Socioeconomics on Risk Perception

The common sight of individuals hanging from the exterior of buses in Mexico City provides a compelling illustration of the social construction of acceptable risk. While objectively hazardous, this practice is tolerated due to a tacit social and economic understanding that these individuals likely cannot afford bus fare. This situation is perceived as a fairer social solution than forcing them to walk long distances. Attempts to alter this behavior by presenting accident or fatality statistics would likely be futile because the social construction of the situation prioritizes economic survival over abstract safety rules and points to the complex nature of risk in modern society.

DOI: 10.4324/9781003668770-8

Table 8.1 Six principles of social construction

Principle	*Meaning*
1. Understanding of Risk	Risk is subjectively perceived and interpreted by individuals based on their experiences, social interactions, and cultural context.
2. Communication & Shared Meaning-Making:	Clear and effective communication is crucial for creating a shared understanding of risk.
3. Multiple Realities & Situational Awareness	There is no single, objective reality. Individuals and groups may interpret the same situation differently. Knowing what is going on, so you can figure out what to do next.
4. Power Dynamics	Power dynamics influence how risk is perceived and managed.
5. Knowledge & Action	Our understanding of risk shapes our actions, and our actions shape our understanding.
6. History and Culture	The historical and cultural context in which safety practices are embedded influences how risk is perceived and managed.

It is common for acceptable risk to vary significantly depending on one's social position and economic circumstances. A person that is starving will be willing to take far more risk to get food than a person who is full.

In this chapter, we analyze the social construction of risk using a framework borrowed from social work: the Micro, Meso, and Macro model (Figure 8.1). Practitioners of this model believe that a person's challenges are rarely just "personal" failures. Instead, they see them as the result of complex interactions between an individual, the groups, and systems around them (families, communities, laws).

These three tiers allow social workers to target interventions at the most effective level—sometimes helping the person directly, and sometimes addressing the system that is hurting them. Social workers rarely stay in just one lane. They use this framework to "zoom in" and "zoom out" to solve a problem effectively.

At the macro level, factors like socioeconomic status, race, and gender roles create significant disparities in safety outcomes. These are not random; they are determined by social structures. As noted by the CDC (2023), marginalized groups often face increased risk exposure and vulnerability to injury, demonstrating how safety is more than a technical issue. It is a reflection of larger societal power dynamics.

Societal forces such as racism, poverty, and gender roles cross the factory gate and increase the odds for who gets (Nicols, 1997).

Socioeconomic Conditions Tied to Higher Fatality Rates

Although the economy isn't within occupational health and safety (OHS) control, an awareness of how the economy affects fatalities may help to enhance empathy. To begin, the stagnating fatality rates indicate a failure to understand the problem

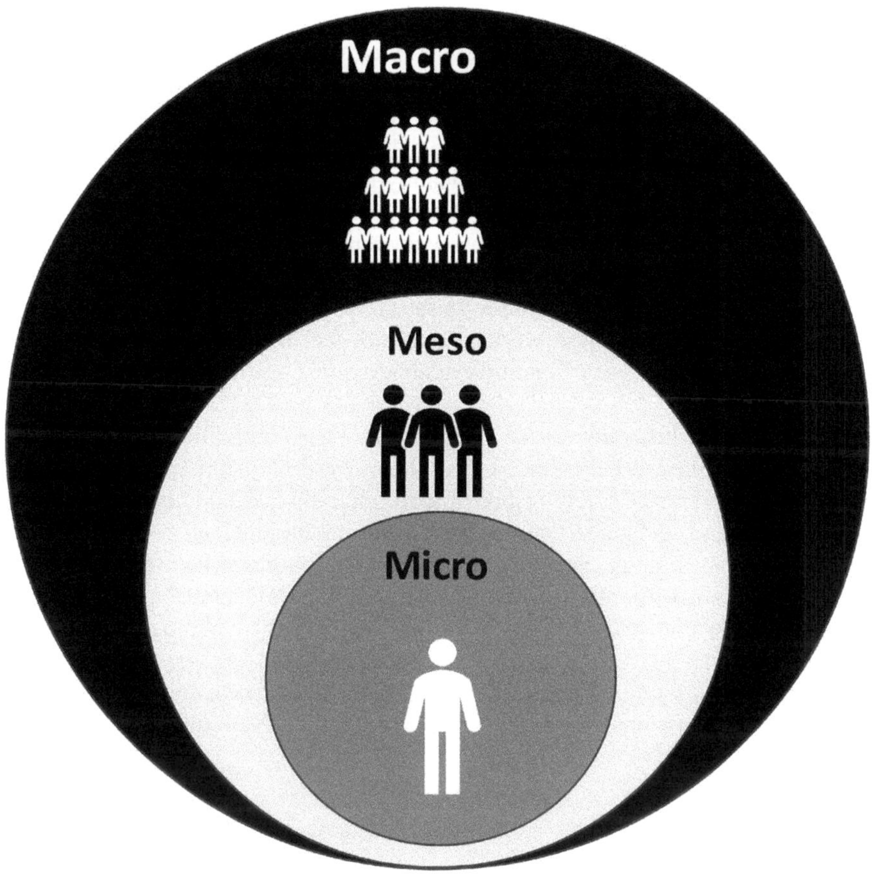

Figure 8.1 Intervention levels.

we are trying to solve. The areas with the highest fatality rates cannot be explained solely by analyzing the safety management systems. There is a deeper story that demonstrates there are social forces at play that prevent the improvement of the safety management systems when they aren't addressed.

Societal factors are not random; they are structures that determine who is exposed to hazards. Research consistently highlights that risk is unequally distributed based on class and race. According to the Bureau of Labor Statistics (2023), the fatal work injury rate for Black or African American workers was 4.2 per 100,000 full-time equivalent workers, and for Hispanic or Latino workers, it was 4.6 per 100,000. Both figures are significantly higher than the all-worker average of 3.7. These numbers are not merely statistical anomalies; they represent systemic inequalities that limit access to safer jobs, adequate training, and proper safety equipment (AFL-CIO, 2024).

Income Inequality

Workers from lower socioeconomic levels often experience higher fatality rates because societal standards implicitly place less value on their well-being. This constructs a reality where their safety is less prioritized, and their injuries are more likely to go unreported due to fear of financial retaliation.

Workers in low-wage jobs are significantly more likely to be injured or killed on the job compared to those in higher-paying positions. Low-wage workers often have less job security and may be more vulnerable to exploitation. They may be less likely to receive adequate safety training or be provided with proper safety equipment. They may also be afraid to raise safety concerns for fear of being fired or otherwise penalized. These workers may not feel they have a choice but to accept unsafe conditions as a condition of employment, even when they are aware of the dangers.

The higher fatality rates also illustrate how social conditions can pressure workers toward unsafe behaviors, even when aware of the risks, contradicting the notion of workers as solely responsible for their choices and highlighting how organizational and societal structures can constrain individual agency.

Hence the psychology or behavior of individual workers has often been the focus of attention, while the social and economic determinants of accidents have been played down or ignored. The inclusion of social and psychological factors in safety would mean addressing power imbalances that limit workers' ability to protect themselves. It would mean expanding risk management to consider the interplay of individual actions, organizational factors, and societal influences (Nichols, 1997;AFL-CIO, 2024).

Racial Disparities

A report by the AFL-CIO (2024) found that Black and Hispanic workers face disproportionately higher fatality rates compared to white workers, particularly in industries like construction and agriculture. This disparity is often linked to systemic social and economic inequalities that limit access to safer jobs, adequate training, and proper safety equipment, showcasing how societal structures construct differential risk.

Black workers experience a higher fatality rate than white workers. Hispanic workers face even greater risk, with a fatality rate exceeding that of both white and Black workers. Systemic inequalities can limit access to better-paying, safer jobs for Black and Hispanic workers. They may be concentrated in more dangerous industries or roles within those industries. As discussed in Chapter 4 when people feel it is too risky to bring up safety issues, it contributes to those higher numbers.

The same AFL-CIO report also estimates that a significant number of workplace injuries go unreported, particularly among workers in precarious employment situations who often come from lower socioeconomic backgrounds. Fear of retaliation, financial penalties, and increased scrutiny contribute to this underreporting, illustrating how social and economic forces construct barriers to accurate safety data.

Industry Concentration

Fatalities are disproportionately concentrated in certain industries and occupations. Construction, transportation, and warehousing consistently report the highest rates of fatal injuries. These industries often have tight deadlines, demanding workloads, and a culture that prioritizes efficiency over safety. Workers might feel pressured to take shortcuts, skip safety steps, or work when fatigued to meet these demands. Challenging these norms, or even suggesting safer alternatives, can be difficult or even career-threatening, making conformity the path of least resistance.

A North Carolina study (UNC, 2024) found that race is a major driver of mortality rate disparities in more dangerous industries. Latino workers in construction and Black workers in food manufacturing face disproportionately higher risks, highlighting the impact of segregation and occupational segregation (social and economic constructs) on workplace fatalities. These findings support the theory of social construction in risk exposure. Workers from lower socioeconomic levels experience higher fatality rates because societies' standards, often implicitly or explicitly, place less value on their well-being, constructing a reality where their safety is less prioritized.

Classism

A concrete example of how socioeconomic disparities show up was during a culture assessment at a nuclear power plant. Employee perceptions of being treated with respect are correlated to organizational performance, including safety (Mansor et al., 2023; Pendell, 2025). In this assessment, there was a stark difference in the perception of being treated with dignity and respect between knowledge and hourly workers. Eighty-four percent of knowledge workers felt respected, whereas only 47% of hourly workers felt the same. This reveals a socially constructed hierarchy of value within the organization. A discussion with the leadership team had little result. A manager told me that even if it's true, no one believes it can change due to an entrenched "us versus them" history.

The irony is that trust and respect are foundational to a nuclear safety culture (INPO, 2004). There were a lot of posters on the walls declaring these espoused values. But only a few understood the connection between their actions and the impact on employees. This cognitive dissonance, where workers are told they are valued but perceive that they are disrespected, aligns with the myth of a level playing field in individualistic societies and mirrors the broader societal undervaluing of certain socioeconomic groups (Kim et al., 2014). This lack of perceived respect directly undermines the trust and communication necessary for positive outcomes, ultimately contributing to the very conditions that can lead to higher rates of serious injuries and fatalities (SIFs) among the most vulnerable segments of the workforce.

In these high-pressure environments social silence, the collective agreement to withhold information about hazards to maintain harmony or prevent job loss, becomes a significant barrier to accurate risk perception. The takeaway here is that survey results should be taken seriously when employees do not feel valued. Perception of mistrust translates into disengagement, hiding mistakes, and a failure to report hazards. Assumptions about national norms cannot be taken seriously by an organization that strives to lower fatalities.

Gender Discrimination: The Social Construction of Risk and Silence

In one-on-one interviews with 56 OHS advisors, half reported that they felt burned out and ready to quit. Many of the women saw themselves in traditionally male-dominated industries like construction encountering discrimination, harassment, and violence due to challenging gender norms. In interviews, they described the stress, depression, and sometimes physical illnesses related to these experiences (Carrillo, 2024).

Women report many more challenges. For example, they are assaulted 36% more often than men (BLS, 2024). They are exposed to more risks in the workplace and get slow responses to concerns about chemical use during pregnancy. One focus group participant said she did not get a response from the OHS manager for eight months (Carrillo, 2012).

One OHS manager described it thus:

> I once had a very smart and intelligent engineer sit in my office during the last week with the organization open up to me that the reason, she was leaving was her opinions and perspective did not matter in a male dominated industry such as mining. Management did not appreciate her response to events, sometimes saying she takes too long to draw conclusions. I wonder how much we are missing by not taking account or perspective of women especially in risk industries. When I was in the oil and gas industry, my crew was known to have fewer events not because we were the smartest but because as a leader who was privileged and honored to have two women on my crew, I leaned on them for advice and made one my Job Site Safety Leader. She was able to see things from a different perspective that we "hard strong macho working" men often missed.

Double Jeopardy: Intersectional Discrimination and Vigilance-as-Resilience

The social construction of risk is perhaps most acutely visible at the intersection of gender and race, creating a double jeopardy of exclusion and devaluation. For women of color in safety-critical roles, the combination of gendered devaluation and racial bias creates profound difficulty in being heard or valued. One Black woman OHS advisor, for instance, was subjected to exclusion, micromanagement, and constant tracking of her whereabouts by her manager, leading her to feel compelled to look for alternatives. She moved on to start her own successful company.

This constant navigation of subtle hostility—microaggressions, exclusion, and systemic disrespect—comes at an emotional cost. It creates hyper-vigilance, a survival mechanism that acutely influences risk awareness. The need to be a "chameleon" to build social relationships and meld into an environment forges a perceptive, resilient, and creative mind to survive.

Hypervigilance-as-Resilience

The constant sense of being at social risk (discrimination) increases an individual's awareness of what is going on around them. It involves more than being aware of your surroundings. It is also making sense of what those elements mean and projecting how they might affect personal or work-related vulnerability. This includes one's personal emotional and physical safety.

This ability becomes hypervigilance for people who grew up and continue to navigate hostile environments. It is an extreme form of social perception, the non-technical ability to perceive and understand the intentions and emotions of others. As one safety advisor recounts, linking this skill directly to their upbringing:

> I am a very hyper vigilant individual due to how I was raised. Growing up in South America, taking risks is a very intense calculated emotional endeavor because if you fail, you are met with physical punishment and the outcome does not just affect you, but affects everyone. Hence my internal system is wired to scan my environment not for the most likely outcomes, but for the most dangerous ones. My mind doesn't stop at the first, easily manageable contingency; it races ahead to the third, fourth, and fifth domino effect–the true cascading failure. Hence as an operator, especially in an environment I was not familiar with, I tend to ask deeper questions than most of my colleagues.

This individual's ability to "race ahead" to the cascading failure is not paranoia; it is *Vigilance-as-Resilience*, a necessary skill honed by experience, which directly affects the ability to recognize anomaly. Anomaly detection is the capacity to recognize when the system is delivering the unexpected and the flexibility to adapt to that condition (Pupulidy, 2015). It is an invaluable source of insight into systemic and non-obvious dangers and how we make sense of these situations.

The irony, and the central tension of this book, is that the very heightened awareness that allows these individuals to detect weak signals of risk is often paired with lower social status, which makes speaking up risky. It is a huge loss because such organizations miss out on the benefits of having people who can detect weak signals.

The advisor's story continues, providing a blueprint for effective leadership:

> Luckily, I had a supervisor who was invested in me and created a space where I felt safe to ask and express all my concerns which were funny sometimes thinking back but it allowed me to truly understand my environment and the discussions around my questions often afforded the crew to discuss their perspectives on a job.

This supervisor demonstrated the concept of *psychological containment*. They did not dismiss the concerns but accepted the individual's anxiety, legitimized the underlying fears, and transformed them into a valuable group discussion, thus making the operator's internal risk model a resource for the entire team. These concepts all come together in the next chapter, describing an organization as a living system with people playing the role of being risk sensors at networks of communication.

Sensemaking

Sensemaking, as defined by Karl Weick (1988), is a collective activity essential for survival. It is a crucial collective activity for survival and the dynamic process by which individuals and groups collaboratively construct meaning from ambiguous and complex experiences. His nuanced approach sought to understand not only what happened but also use that information to manage the unexpected (Weick et al., 2015).

Weick is likely one of the people most responsible for moving safety investigations away from finding blame toward understanding the factors that made the operator's actions seem reasonable, rational, or even necessary at the time. However, while sensemaking is accepted as an explanation of how people come to consensus on what is true and what to do, it is criticized for failing to adequately account for the roles of power, politics, and culture in shaping what people ultimately decide what to do (Bridger, 2021). This radical reframing is not merely an academic exercise; it is the central question that drives the innovative, learning-focused accident review methodologies that have emerged over the past decade (Turner & Gray, 2009; Bandura, 1977).

The Learning Review (LR) developed by Ivan Pupulidy (2015) and implemented by the US Forest Service stands as a transformative example of sensemaking theory put into practice. It was explicitly designed to replace the traditional investigations, which were found to be ineffective at preventing fatalities and often resulted in blaming firefighters for their own deaths. The core principle of the LR is to shift the focus from investigation to inquiry. It moves away from judging actions as right or wrong and instead seeks to understand the context surrounding human decisions by asking: "Why did it make sense for them to do what they did at that time?"

In short, sensemaking is the active, ongoing process individuals and groups use to create shared understandings known as social construction.

Sensemaking in Incident Investigations

Traditional investigation models such as the "5 Whys," Fishbone diagrams, or TapRooT are rooted in a mechanical worldview. They assume that accidents are linear sequences where one domino knocks over another, eventually leading to a "Root Cause," usually a broken component or *human error*. However, the very concept of a "Root Cause" is a myth because complex social systems do not have linear, single origins. By stripping away the "messy" context to find a single cause, traditional investigations often create a "Retributive Culture" focused on finding fault rather than achieving understanding (Long, 2012).

When the root cause is found to be human error, investigators effectively end the inquiry exactly where it should begin. When we invoke personal responsibility as the cause of a failure, we treat the worker as a broken component to be fixed or replaced, rather than a signal that the system is out of alignment.

Risk Makes Sense

The foundational premise is simple yet transformative: Risk makes sense to the person doing it at the time. Skilled workers do not go to work intending to harm themselves. Therefore, if an employee bypasses a safety interlock or enters a confined space without a permit, that decision was not "stupid" or "irrational" in their immediate reality. It was a decision shaped by heuristics, production pressures, and cultural norms that made the risk appear to be the most viable option.

Ivan Pupulidy, creator of the LR for the US Forest Service, argues that we must shift the investigative lens from judging behavior to understanding context. Instead of asking, "Why did they violate the rule?", the LR asks, "Why did it make sense for them to do what they did?" This question primes the investigation to look for performance-shaping factors—such as conflicting goals or resource constraints—that mainstream methods ignore.

The obsession with data and technology ignores the critical element of human judgment and sensemaking during an event. The focus on statistical analysis and the search for a root cause actively discourages the nuanced, contextual understanding that is vital for effective learning. The plethora of data ultimately avoids the most critical aspect of investigation: the complex interplay of human interactions, judgment, and decision-making within a specific socio-cultural context. By not grappling with the experiences and perspectives of the individuals involved, very little can be learned that would prevent another event. This approach locks us into a limited learning framework designed to find and fix but not to understand. This approach does not address deep underlying assumptions—it accepts these assumptions as truths and the system remains unchanged (Pupulidy, 2015).

From Interrogation to Inquiry

Traditional investigations ask: "Why did you violate the rule?" This creates defensiveness. A social constructionist approach asks: "Why did it make sense to do what you did?" This approach builds trust, reveals the work as actually done, and exposes the systemic trade-offs workers are forced to make every day.

Dr. Nippin Anand, an experienced accident investigator, provides the practical language for this shift. In a traditional investigation, a supervisor might ask a worker who dropped a load, "Did you follow the procedure?" This is a closed question that invites a defensive "Yes/No" response and focuses on compliance (Anand, 2024).

Anand gives this example: A worker almost dropped a load with his forklift. In the old way, the supervisor would focus on rules broken or mistakes made, probably leading to the worker getting blamed. A better, human-centered approach shifts the focus to understanding the worker's experience, their state of mind, what motivated them, who they work with, and any hidden biases. So the supervisor asks, "Tell me about your experience using this equipment. How were you trained, and what have you learned on your own?" This human-centered approach shifts the focus to the worker's lived experience

Social Construction

Social constructs are shared beliefs, which can reshape reality itself. As an example, one of the earliest and most significant was the reconstruction of a flat earth into a sphere.

Applied to OHS, social construction means that risk is not an objective truth. Risk is defined through ongoing human interactions, agreements, cultural norms, and power relationships within an organization.

Per Pupulidy, "Just as Weick locates sensemaking as happening between people, Cooperrider and Whitney describe constructionism as replacing the individual with relationships. "Constructionism is an approach to human science that replaces the individual with the relationship as the locus of knowledge (Pupulidy, 2015, p. 50)."

"Sensemaking as a function of relationship—can be seen as capacity to understand or make meaning of these relationships. This points to organizations, regardless of size, as sensemaking entities" (Pupidily, 2015: 150).

This shift in perspective suggests that organizations function as a complex network of relationships that generate, contain, and leverage organizational knowledge. Risk is both constantly fluctuating and intangible; therefore, it must be collaboratively defined through social construction. The most important tool of social construction is language.

Sensemaking is an ongoing process where we look back at things that happened and create plausible stories about what happened, what we heard or saw. When it is something, we've never seen before we must make sense of it, turning the unknown into something we can comprehend and act upon (Weick, 1988).

Sensemaking influences risk perception, rule interpretation, and decision-making. To truly learn from disasters, we need to shift our focus to a collective understanding of risk. As Karl Weick insightfully suggested, "Human errors are fundamentally 'caused' by human variability, which cannot be designed away … Errors are less likely to enlarge if they are understood more fully, more quickly." This necessitates involving individuals to collectively make sense of risk within complex systems and how their social dynamics can either amplify or mitigate errors (Weick, 1988: 308).

Weick also proposed that this form of sensemaking can form the basis for an ideology of crisis prevention, where a "relatively coherent set of beliefs" binds people together and guides their actions. Integrating regular discussions to make sense of risk in work planning can help identify and address the deviations from procedure, promoting a more conscious and informed approach to safety. This practice enables people without formal power to actively participate in shaping what "good safety" looks like in S&H (Mathieu et al., 2000). The following tools have proven to prepare people for sensemaking and collaboration around identifying risk.

Sensemaking Conversations

"An organization is a conversation" (Broekstra, 1998: 175). This profound statement posits that an enterprise exists solely within the relationships it fosters. As Stacey (2001) argues, there is no organization outside of these human connections. If the flow of communication regarding work and purpose ceases, the organization effectively ceases to exist. It is this continuous interaction—what we call sensemaking—that brings organizational reality into being.

Contrary to the mechanistic view of a company as a collection of gears and levers, it is a living entity, sustained by sharing and discussion of ideas. In this way the organization is continually reproduced and reshaped by a vast network of micro-interactions.

Sensemaking turns talk into structure, making it a potent tool for managing risk. Perceptions of risk shift only when you alter the conversations people use to interpret their work (Kramer, 2016). Consequently, Shaw (2002) argues that true leadership requires stepping into the "messy" exchanges where this reality is negotiated in real-time.

Conversation as a Structural Sensemaking Tool for Safety

If people are not talking about risk, it is not being managed. On the other hand, if they don't understand the risk they are talking about, that is dangerous as well. In *The safety Counterinsurgency,* Brooks, a safety manager, describes an alarming event that radically changed their safety program.

> Two employees, unsure of which valve to turn during a critical operation, decided to flip a coin to choose. When Brooks discovered this, he realized that the existing Stop Work Authority (SWA) policy was a failure. The workers felt it was safer to gamble on a coin toss than to stop the job and ask for help.

That story is a great example of how the quality of an organization is defined by the quality of its conversations (Flores, 2012). A high-value exchange is not merely a transfer of data; it is a generative act. It constructs a shared reality, resolves uncertainty, and coordinates effective action. It is the distance between a slogan, "Safety First" and a binding commitment "I will place the barricade before I start."

These generative interactions allow the "messy truth" to surface without damaging relationships. They validate the emotions and lived experiences of participants, establishing the psychological safety required for candor. In the safety context, this distinguishes the feeling of an audit (policing) from the experience of fieldwork (partnering). The ultimate litmus test for this quality is simple: Did we leave the interaction with a different understanding of the risk than when we entered?

Conversation is a powerful catalyst for change. It is the only vehicle for change other than coercion (Stacey, 2001; Schein & Schein, 2018; Carrillo, 2019; Shaw, 2002). It represents the most frequent and accessible opportunity to influence outcomes.

If safety is indeed a social construction, then improving risk management requires nothing less than intervening in the conversations that construct that reality.

Organizational and Social Constraints on Conversation

Belief #1: "True communication takes place in the presence of relationship and trust," is best achieved through conversation. Yet, most managers and executives make limited time for them. A lack of understanding about the nature of conversation contributes to this perception. Consequently, emails are the most frequently used communication, contributing to frequent misunderstandings.

If you have ever tried to engage a group of operational employees in sensemaking, you may have experienced resistance. Maybe they complained that it took too long and was a waste of time. Or they may have felt that it was simply an academic exercise. That is the reality in the workplace.

For me, the answer to them is in the work of Paul Slovic (2000). His work proves that what truly drives action when facing danger is the emotional response to risk—the dread, the sense of control, or the feeling of trust in the system. So, unless a team spends time before the crisis developing a shared understanding and vocabulary around how to handle a risk, they are unlikely to figure out the correct action to an unexpected risk. That may be what happened in the Mann Gulch firefight, as discussed in the analysis in Chapter 6. The team leader survived because he could sensemake a solution, but he could not convince his team to follow him. Their fear caused them to run away. Consequently, everyone else on the team died except him. Unfortunately, they had not discussed that potential scenario prior to the firefight.

This story is a cautionary tale because sensemaking in a crisis will not work if you do not have a relationship with the people involved prior to the event. And it is important to practice those scenarios repeatedly.

Fear of Social Interaction Stops Conversation

Besides time, fear of social interaction plays a role. Each time we ask for or give information there is a risk of embarrassment or rejection. This has to do with the fear of losing face. People admit they don't ask questions because they don't want to appear incompetent. Research shows managers fear failure, public embarrassment, or board/market reactions, which can lead them to suppress bad news or resist uncomfortable changes (Kegan & Lahey, 2016). Managers are often criticized for walking past unsafe behavior with the assumption that they don't value safety. But this judgment fails to consider the complex social dynamics at play. This behavior is often deeply intertwined with the desire for saving face—for both the employee and the manager. Avoiding public shaming to maintain a positive relationship can be prioritized over safety vigilance. Freeing people from this fear unleashes human potential to share ideas, collaborate, and watch out for each other in the face of danger.

The fear and the benefit of letting it go were confirmed by a nuclear safety advisor with 30 years of experience who shared:

> I've done scores of walkarounds/worker interactions with managers and it's true many seem intimidated (i.e., fearful) of their own people. Or, perhaps it is a fear of looking weak. Some managers are, of course, natural communicators and enjoy the interactions. I had a plant manager who liked to go to the plant on back shifts to interact with his people. He often asked me to go along and when we showed up at 2:00 AM I felt we were treated almost like rock stars. The appreciation that the site's top manager cared about his people and what they were doing was palpable.
>
> (Personal correspondence)

What People Value in Leaders

Another safety advisor suggests that the hesitation to interact in conversations might stem from a misunderstanding of what people value in a leader:

> I know that, for example, on oil rigs people are promoted primarily for their knowledge and experience yet the most respected leaders funnily enough were never popular for those things. They were always liked because they knew how to treat people. I understand technical knowledge is important in lots of roles but sometimes too much emphasis goes on that and despite all this knowledge people never wanted to be led by them and lots of things fell by the wayside because of it. I suppose getting that balance is tough but it is essential to consider it and try to cultivate it, as technical alone or too big an imbalance just doesn't work in my mind yet I've seen this in action too often.
>
> (Personal correspondence with HSE Advisor in offshore drilling and pipeline modification for 10 years)

These stories illustrate both the value that employees place on conversations and the self-imposed or external barriers holding them back. A manager or safety practitioner who cannot go into the field and comfortably ask or answer questions is at a great disadvantage. They will not have access to accurate information about what is going on at the front lines, whether it is in the office or the field. There are many creative ways to do this and they are shared in Chapter 11, The Counterinsurgency.

Sensemaking and Social Construction

Sensemaking and social construction are interconnected processes. Sensemaking involves individuals interpreting ambiguous or uncertain situations and information, while social constructionism highlights how shared meanings are developed and sustained through dialogue and other social interactions. These processes reinforce each other: individual interpretations from sense-making contribute to the social construction of reality, and conversely, social interactions influence an individual's sensemaking.

Many of the mistakes we've made when handling an issue or communicating about safety were based on faulty sensemaking that became part of our reality (social construction). In practical terms and in hindsight our mistakes can be traced back to a lack of knowledge, acting on unquestioned experience, or biases.

If the mistake was caused by a lack of knowledge we can gain more knowledge, but there are always unexpected circumstances. It is harder to correct biases because they are often unconscious, and secondly because bias is a human characteristic that can only be managed, not eliminated. So, our best course of action is to improve the way we listen and ask questions.

The Social Construction of Risk

The concept of organizations as conversations implies that organizational outcomes are fundamentally shaped by the quality of dialogue and interaction. This would mean that the level of performance you are experiencing in your organization is a result of the same things as what your people are discussing.

Are your people discussing safety risks or schedule and cost? Are they talking about risks in a knowledgeable way? Are people exchanging relevant information in a timely manner? The only way to answer these questions is to be in conversations with the people you work with. Of course, the effectiveness of these exchanges depends on the trust levels. When discussing this topic with an OHS leader, they observed:

> This to me is one of the most important traits any safety leader or leader in general needs—to develop trusting relationships. This has been the secret to my success as a safety leader. Creating connections and being able to sit with employees expressing their true feelings about the organization has led to a successful relationship with these miners. They now call me to ask for clarifications and my opinion before doing certain things.
>
> (Private communication with OHS advisor)

To succeed, there must be a relationship with a basic trust level for people to be open about the way they do things to construct a working understanding of acceptable risk. Analyzing these deviations reveals deeply ingrained social constructs about what is truly "acceptable" versus what is formally prescribed.

Case Study: Application of Social Construction to Rail Crash

Chikudate (2009) provides a stark example of how a culture of individual blame can undermine how people view safety. His analysis of a devastating 2005 railway crash in Japan (107 fatalities, 562 injuries), documented in "If human errors are assumed as crimes in a safety culture: A lifeworld analysis of a rail crash," challenges the official reports that primarily blamed the driver. The article meticulously documents how an emphasis on shame and embarrassment shaped worker identities and led to a climate of fear, discouraging open discussion to achieve a shared understanding of risk and safety.

His research suggests that the driver's decision to speed cannot be fully understood without considering how the prevailing safety culture and learning practices, emphasizing individual accountability and shame, created conditions where the socially constructed reality made adhering to the schedule without speeding perceived as extremely difficult, if not practically impossible. His analysis of organizational culture can guide efforts to create a more just and open reporting system.

Flores (2012) from Chile, also from a collective culture, articulates the process by which accountability is created via one structured conversation after another. It cannot be forged by coercion. He is now considered the father of a branch of coaching called ontological coaching, which uses language as a way to change one's reality.

Case Study: Firefighters

The Incident Command System developed by the US Forest Service helps the firefighters and partners manage complex incidents by providing guidance to allow the incorporation of inputs from all levels of personnel engaged in an emergency response.

It was found that improved organizational safety arises from the way individuals relate with each other during unfolding events. Evidence from this study suggests that questioning (providing upward voice) becomes a critical driver of safety insofar as it triggers sensemaking and re-evaluation. The individual's level and ability to participate depend on feeling safe from any repercussions.

Ladder of Inference as a Sensemaking Tool

How do we improve sensemaking in the daily grind of business? We must slow down our thinking processes. The Ladder of Inference is a useful tool to demonstrate how our assumptions and mental models can contribute to the "messiness" of communication. Chris Argyris (1974) introduced the Ladder of Inference (Figure 8.2) to demonstrate how people individually make sense of what they see and hear, and how two people observing or hearing the same thing can end up with a completely different understanding of what happened. It then explains how individuals make decisions and take action. As individuals understand this process, they can use this knowledge to construct a shared understanding of an event. In other words, it improves sensemaking capabilities.

Argyris proposed that people are not passive observers of an objective reality. Through their actions, they create the very environment they then interpret. For example, in a crisis, actions taken to understand the situation—flipping a switch, opening a valve, running a diagnostic—simultaneously alter the situation itself generating new data that must then be made sense of. Weick also called this phenomenon enactment (1988).

Ladder of Inference Example: Negative Context

1 Observable data: People working.
2 I select data: We see a worker without gloves.
3 I affix meaning: Worker is violating safety rule.
4 I make assumptions: Someone is willfully breaking the rules.
5 I draw conclusions: They are negligent.
6 I draw upon my beliefs: If I don't make them an example, others will think they can skip protective equipment.
7 I take action: Take disciplinary action.

By understanding this ladder, leaders can stop at the "Assumption" rung and ask: "Wait, what data am I selecting? What meaning am I adding?" This tool transforms a snap judgment into a sensemaking conversation.

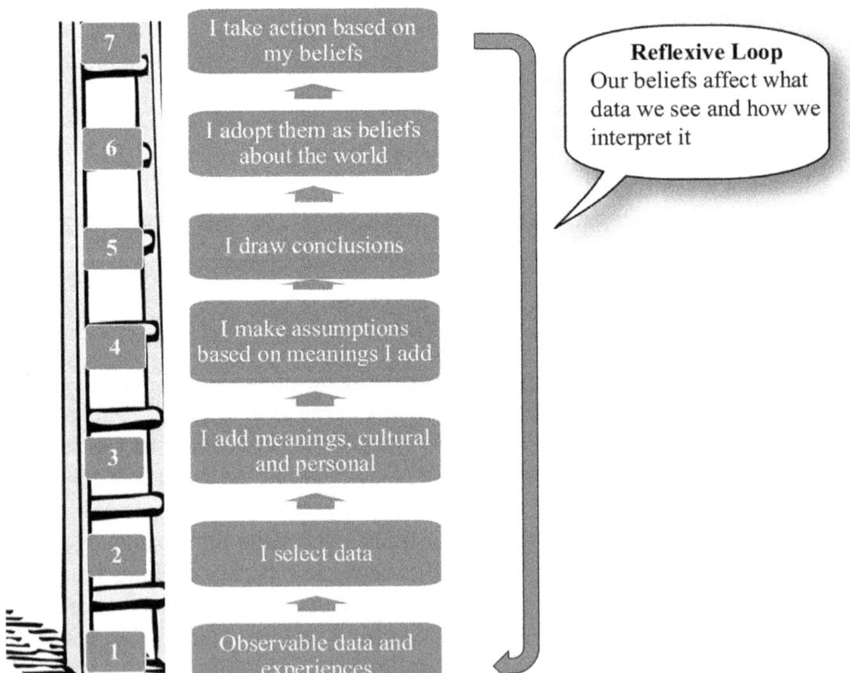

Figure 8.2 Ladder of Inference.

In a complex environment with a lot going on, people are taking multiple actions and extracting specific cues to build their understanding. What is noticed and what is ignored depend on context, experience, and expectations, which could also be biased. The ladder is a tool that can be used for training under different scenarios to practice sensemaking and learn how to manage the art of questioning assumptions.

Neuroscience has advanced significantly since Argyris' ladder. For example, he did not call out that emotion preceded action. Emotions are crucial for decision-making (Damasio, 1994). An idea alone is insufficient to drive action; an emotional connection is necessary. What could that be for people in the workplace? As an example, Weick recounts that the months before the Bhopal disaster, management demonstrated a collective sense of indifference and a disregard for safety, having decided that the plant was closing and no longer of value. The emotional reaction of the workers was to stop paying attention to the equipment and a loss of motivation to notice or report anything. As Weick states, "Top management's perceptions of Bhopal set into motion enactments that confirmed their perceptions. It became a self-fulfilling prophecy" (1988: 313).

A leader can use this model with their team to arrive at a shared understanding of a problem or an event. A skilled facilitator can learn to guide this process with real-time safety issues.

Increased Situational Awareness

Why is this so important? There are millions of bits of sensory information coming at you every second, but you're only aware of a tiny fraction. What determines which information gets through? It is whatever you've trained your brain to value and notice. That's why intention matters so much. Where you put your attention, it grows. Where you take it away, it withers.

Understanding how we interpret what we see and hear has profound implications. It helps us move beyond fight-or-flight responses, revealing more opportunities for action. This awareness reduces unnecessary anxiety stemming from misinterpreted intentions and allows us to remain objective, fostering inquiry over premature judgment. Essentially, learning to listen non-judgmentally, empathetically, and with curiosity enhances our effectiveness and reduces mental stress.

(A facilitation guide for the Ladder of Inference is available in Appendix D.)

Conclusion

This chapter redefines risk management, moving it from a purely technical exercise to a strategic social activity. For an executive, the message is clear: your organization's true risk profile is not determined by procedures on paper but by the shared understandings, conversations, and power dynamics that shape your employees' daily reality. Ignoring these social forces is a critical business oversight that directly contributes to flawed risk assessments and a failure to reduce SIFs (Barton, 2024).

Executive Summary

The performance your organization is experiencing today is a direct reflection of the conversations your people are having. True risk management requires leaders to move beyond the illusion of control offered by traditional systems and engage directly in these conversations.

By embracing the fact that risk is a social activity, you can shift your organization from a reactive, compliance-driven posture to a proactive, human-centered one. This fosters a more resilient culture where trust, open communication, and collective sensemaking become your most powerful defenses against catastrophe. This is not just about preventing SIFs; it is about building a more engaged, adaptive, and high-performing organization.

Takeaways

- *Redefine Your Understanding of Risk*: Risk is subjective and shaped by experience. Involve frontline workers directly in risk assessment processes. Go beyond checklists to actively understand how they perceive hazards in their specific environment. The Opticore Tech "ride alongs" are an excellent model for this.
- *Master Communication to Build Shared Meaning*: Clear, inclusive language is essential. Create safe spaces for open dialogue where workers can voice concerns and challenge practices without fear of reprisal. This builds the trust necessary for true communication.
- *Acknowledge Multiple Realities*: Different groups interpret situations differently. Actively encourage and facilitate perspective-sharing across teams and hierarchical levels. Use sensemaking questions like: "What do they know that I need to know? What do I know that they need to know?" to build a complete operational picture.
- *Manage Power Dynamics Intentionally*: Power influences how risk is perceived and managed. Empower workers by giving them a genuine voice in safety decisions. This is the essence of RCS Belief #2: Inclusion precedes accountability.
- *Link Knowledge to Action*: How we understand risk shapes how we act. Treat mistakes and deviations from procedure as invaluable opportunities for collective learning, not just as moments for punishment.
- *Adapt to History and Culture*: Safety practices must fit the cultural context. Recognize that a "one-size-fits-all" safety program is a strategic risk. Tailor communication and strategies to be sensitive to the cultural backgrounds of your workforce, especially in global operations.

References

AFL-CIO. (2024, April 25). *Death on the Job: The Toll of Neglect*, 2024 (33rd ed.). https://aflcio.org/reports/death-job-toll-neglect-2024.

Anand, N. (2024). *Are We Learning from Accidents?* Novellus. https://nippinanand.com/wp-content/uploads/2024/04/Are_We_Learning_from_Accidents_sample.pdf

Argyris, C., & Schön, D. (1974). *Theory in Practice: Increasing Professional Effectiveness.* Josey Bass.

Barton, M. A. (2024, April 25). *Death on the Job: The Toll of Neglect* (33rd ed.). https://aflcio.org/reports/death-job-toll-neglect-2024

BLS. (2024). Fatalities. www.bls.gov/news.release/cfoi.nr0.htm

Bridger, R. S. (2021). *Sense Making in Safety Critical and Complex Situations.* Routledge. Chapter 2. A guide to human factors in accident investigations. www.researchgate.net/publication/362950356_A_Guide_to_Human_Factors_in_Accident_Investigation/fulltext/6308cc725eed5e4bd11f7d87/A-Guide-to-Human-Factors-in-Accident-Investigation.pd

Broekstra's, G. (1998). An organization is a conversation. In: Grant, D., Keenoy, T., & Swick, C. (Eds). *Discourse and Organization.* Sage Publications, Inc.

Bureau of Labor Statistics. (2023, December 19). *National Census of Fatal Occupational Injuries* in 2022 (USDL-23-2615) [News release]. U.S. Department of Labor. www.bls.gov/news.release/pdf/cfoi.pdf

Carrillo, R. A. (2012). Relationship-based safety. *Professional Safety Journal, 57*(12): 35–45.

Carrillo, R. A. (2019). *The Relationship Factor in Safety Leadership: Achieving Success through Employee Engagement.* Routledge.

Carrillo, R. A. (2024). *OHS Voices from the resistance.* https://carrilloconsultants.com

CDC. (2023). Sexual Violence Prevention. https://www.cdc.gov/sexual-violence/about/?CDC_AAref_Val=https://www.cdc.gov/violenceprevention/sexualviolence/fastfact.html#content

Chikudate, N. (2009). If human errors are assumed as crimes in a safety culture: A lifeworld analysis of a rail crash. *Human Relations*, 62: 1267–1287.

Damasio, A. (1994). *Descartes' Error: Emotion, Reason, and the Human Brain.* Penguin Books.

Flores, F. (2012). *Conversations for Action and Collected Essays: Instilling a Culture of Commitment in Working Relationships.* Self published.

Horton, C. (2024). The silent strain: How debt takes a toll on mental health. *Forbes Advisor.* www.forbes.com/advisor/banking/american-debt-and-the-mental-health-epidemic/#:~:text=How%20Debt%20Impacts%20Mental%20Health,stressed%20because%20of%20their%20debt

INPO. (2004). *Principles for a Strong Nuclear Safety Culture.* NRC. www.nrc.gov/docs/ML0534/ML053410342.pdf

Kegan, R. & Lahey, L. L. (2016). *An Everyone Culture: Becoming a Deliberately Developmental Organization.* Harvard Business Review Press.

Kim, S. K., Kim, H., Lee, K., & Kang, H. T. (2014). The relationship between injury and socioeconomic status in reference to the fourth Korean national health and nutrition examination survey. *Occupational and Environmental Medicine*, 26(1): 1. https://doi.org/10.1186/2052-4374-26-1

Long, R., & Long, J. (2012). *Risk Makes Sense.* Human Dymensions.

Mansor, F., Huzaimi, Y., Hashim, M., Muhammad, N., & Omar, N. (2023). Employee engagement and organizational performance. *International Journal of Accounting, Finance and Business (IJAFB)*, 8(50), 69–80.

Mathieu, J. E., Heffner, T. S., Goodwin, G. F., Salas, E., & Cannon-Bowers, J. A. (2000). The influence of shared mental models on team process and performance. *Journal of Applied Psychology*, 85(2): 273–283.

Nichols, T. (1997). *The Sociology of Industrial Injury*. London: Continuum.
Pendell, R. (2025). Respect at work returns to a new low. *Gallup*. www.gallup.com/workplace/655040/respect-work-returns-record-low.aspx
Pupulidy, I. (2015). *The Transformation of Accident Investigation: From Finding Cause to Sensemaking*. Published Dissertation, Tilburg University, Tilburg, The Netherlands.
Schein, E., & Schein, P. (2018). *Humble Leadership*. Berrett-Koehler.
Shaw, P. (2002). *Changing Conversations in Organizations: A Complexity Approach to Change*. Routledge.
Slovic, P. (2000). *The Perception of Risk*. Oxford University Press.
Stacey, R. (2001). *Complex Responsive Processes in Organizations: Learning and Knowledge Creation*. Routledge.
Sutcliffe, K. M., & Barton, M. A. (2009). Overcoming dysfunctional momentum: Organizational safety as a social achievement. *Human Relations*, 62(9): 1327–1356.
Turner, N., & Gray, G. (2009). Socially constructing safety. *Human Relations*, 62(9): 1259–1266. https://doi.org/10.1177/0018726709339863
UNC (North Carolina Health Authorities). (2024). *NC Health Disparities Analysis Report*. www.ncdhhs.gov/2024-health-disparities-analysis-report/download?attachment
Weick, K. (1988). *Enacted sensemaking in crisis situations*. http://wendynorris.com/wp-content/uploads/2018/08/Weick-1988-Enacted-Sensemaking-in-Crisis-Situations.pdf
Weick, K. E., & Sutcliffe, K. M. (2015). *Managing the Unexpected: Sustained Performance in a Complex World*. John Wiley & Sons Inc.

9 A Strategic Blueprint for Relationship-Centered Safety

Introduction

This chapter presents the strategic why for a relationship centered strategy and how to embed these concepts into your organization. It introduces the organization as a living system metaphor, which reframes safety from a cost center into a strategic imperative for organizational health, resilience, and business continuity. So we begin with an example of an organization operating as a living system in ways that everyone said could not be done.

AES, Inc. is a power company that grew from a $1 million in investment to over $8 billion in ten years under a philosophy strikingly like the precepts covered in this book. As described by Bakke, co-founder of AES in his book, *Joy at Work* (2005), from its inception AES set out to eliminate all formal approval mechanisms at the company. This allowed the teams to make decisions at their level about everything from hiring to permission for vacations and budgets. He created an environment where leaders coached decisions more than made them. He was also a big proponent of building relationships, rather than rules. AES was the physical proof that organizations could function as living systems with people networked but not controlled (Wheatley, 1992;1997).

Under Bakke, there were no safety or human resources departments. He felt specialists did not understand operations and operating groups did not understand finance and strategic planning. By eliminating "specialists," teams of people interested in the whole operation were created. During his tenure AES had a better than average safety record (Manz & Sims, 2001).

Bakke also emphasized the promotion of relationships through employee surveys. He read and responded to every comment sent to him through the bi-annual perception surveys. In the beginning, it took one hour to read the comments; toward the end it took about five months to read them because his responsiveness generated so much participation (114).

Perhaps the most controversial practice was the level of spending authority for each employee. Bakke eliminated caps and replaced them with the "advice process." Each decision-maker (person charged with investigating an investment in an idea or initiative) had to consult with knowledgeable peers and senior-level people until

DOI: 10.4324/9781003668770-9

he felt that they were in alignment with the needs of the team, plant, and company. There were famous stories of employees saving the company thousands of dollars by using their credit card to buy needed equipment at a much lower cost than originally quoted by a vendor. There was also a story about an employee who bought the wrong parts for thousands of dollars yet did not suffer negative consequences. These were powerful stories that fueled employee's sense of teamwork.

A 1999 *Harvard Business Review* interview with Dennis Bakke and Roger Sant asked the question of how they handled mistakes. Both agreed that if someone accepted responsibility for the error and apologized then no punishment was needed. They did clarify that after the second or third time action would be taken. They went on to say that the advantage of owning up to your mistakes right away is that it is so much easier to move quickly to find a creative solution. You don't sit around wasting time trying to figure out whom to blame (Wetlaufer, 1999).

AES outperformed its competitors and all worked well until the 2008 financial crisis. The board of directors removed Bakke when energy-related stocks crashed. Nevertheless, from 1981 to 2008 AES had grown to an $8 billion dollar company using principles where people were collectively trusted to make important decisions with their peers without the bureaucracy of corporate control. If it happened once, it is possible for other leaders to duplicate this success.

The impressive growth and ultimate challenges faced by AES highlight the profound impact of empowering people to make decisions at the level where work happens. How did teams self-manage and yet maintain effective communication with the entire organization. This brings us to the theoretical frameworks that offer valuable insights into how such collective trust and shared understanding are built and maintained within an organization.

Two metaphors guide this transition. The first is that of the organization as a living system complete with an immune system. This metaphor reconstructs our understanding of communication as an organic process rather than a rigid, mechanical one. This reframing positions the strategic framework for human-centered safety not as a simple, step-by-step change model, but as a dynamic system. The second metaphor, a Relationship-Centered Safety (RCS) approach, positions an organization as a holistic system where every individual is interconnected. Thus the state of the relationships in the organization determines organizational outcomes. RCS provides the core beliefs and leadership approach to build the strong connections that a resilient immune system requires.

Organizations as Living Systems: A Strategic Blueprint for RCS

Today the speed of change and increasing complexity in the environment have forced organizations to adapt to changing conditions if they want to thrive. That has led to the study of applications of organic living systems to business (Pawar, 2025).

The concept of organizations as living systems is not new. Extensive work, particularly in systems thinking, has brought this perspective into the limelight. By exploring some of the biological models from our human physiology, we can gain insight and generate "new" ideas for managing risk.

The following framework borrows from Margaret Wheatley's concept of collective organizational intelligence and her idea that the speed of learning is directly related to the number of interconnections among the members of that group and with outside entities (2006). This connectivity is embodied in the relationships among the people in an organization or group. Thus, the name *Relationship-Centered Safety*, also known as relationship-based safety, functions as a living system.

Imagine your organization as a living organism. Its overall health and resilience are continuously protected from internal and external threats by its immune system. Within this organism, there is a constant flow of receiving and processing vital information from every *nerve ending* (employee and operational touchpoint), but there is also a communication hub (made up of leaders) to coordinate and direct the immune system's complex responses to maintain the organization's well-being.

Leaders act as a communication hub, interpreting information from *nerve endings*. Each leader processes the information, addresses the threat at their level, and passes the information onto the central communication hub at the executive level. Executives set strategic direction, make crucial decisions about allocating resources to the *immune cells* (employees) to address threats proactively. They are responsible for promoting a resilient culture, encouraging collaboration, and ensuring that all parts of the *immune* system are working together effectively.

This doesn't mean that only *leaders* respond to threats. A healthy immune system is built from the collective wisdom and interactions of every cell—every employee. There is no central lab distributing just-in-time antibodies. It is a *collective* system that constantly monitors for threats, adapts to new pathogens, and coordinates defenses. Examples of pathogens/threats in the organization:

- Process inefficiencies (like a chronic low-grade infection)
- Erosion of trust (weakening of the immune system's general health)
- Market shifts or technological disruptions (new, evolving pathogens)
- Ethical breaches (a severe, acute infection)

***Soft Infrastructure*: Essential Components of the Organizations Living System**

The metaphor of organizations as living systems provides a framework to view an organization as a living system where people play a vital role as components of the immune system. Daniel Katz and Robert Kahn (1978) proposed that organizations as open systems must actively fight against the natural tendency toward decay and disorganization. Thus, safety defense can be viewed as nurturing the organizational immune system—fostering vital internal communication, valuing collective sensemaking, and prioritizing protection over production. As these practices are reinforced, over time, they can develop a shared "mind"—a collective sense or set of beliefs that guide how they act (McShane & Von Glinow, 2021). All these practices help leaders do more than prevent disasters. They are building a measureably healthier, more adaptable, and ultimately, more robust, and sustainable enterprise.

Per the UN, soft infrastructure, encompassing social networks, knowledge, and shared practices, plays a critical role in effective disaster management and risk reduction, according to the United Nations. It complements and enables the functioning of "hard" infrastructure (physical resources like roads and buildings) during and after disasters (Ullberg and Warner, 2016).

In an organization soft infrastructure is akin to the operating system. It encompasses the vital, often intangible, elements that underpin an organization's functioning such as relationships, information sharing, and management practices. People are not merely supportive components in the infrastructure. They function as the essential software that dictates how an organization's formal "hardware"—its departments, hierarchies, and systems—operate.

Without effective soft infrastructure, even the most robust formal structures can falter, leading to inefficiencies, communication breakdowns, and increased vulnerability to unforeseen challenges. This includes elements like trust, shared values, collaborative practices, and open communication channels, which act as the social glue holding an organization together. Investing in and nurturing these intangible assets are essential for fostering resilience, adaptability, and long-term success in a dynamic environment, enabling organizations to navigate complexity and respond effectively to change.

The Ethic of Care

The ethic of care is the moral operating system for a living system, providing the rationale for why nurturing the system's social health is a strategic imperative. It directly supports the concept of Generativity, which serves as the scientific and psychological foundation for a relationship-centered approach. The "ethic of care" is the practical expression of this generative motive—the desire to guide, support, and benefit future generations within the organization (Erikson, 1950, 1982).

This ethic is integrated into RCS. For example, RCS Belief #8, "Relationships influence emotions, feelings and beliefs, which influence decisions," directly reflects an ethic that values empathy and compassion in decision-making. It offers a powerful counterpoint to the individualistic mindset that prioritizes profit over people, a theme explored in cases like Massey Energy and Disneyland, where a lack of care led to fatalities (see Chapters 5 and 7).

According to Carol Gilligan, the ethic of care is a moral position that emphasizes caring relationships, empathy, and compassion in decision-making (2011a). However, the ethic of care is not culturally encouraged for men, which can make it difficult for those who embrace it. Gilligan's concept is considered feminist because she rejects the idea that caring is for the weak—and therefore a feminine characteristic. Most businesses are patriarchal, making the ethic of care an uncomfortable fit. Its adoption is further complicated by its nature as a feminist ethic of resistance to injustices inherent in the patriarchal system.

As Carol Gilligan (2011b) stated:

> A feminist ethic of care guides the historic struggle to free democracy from patriarchy; it is the ethic of a democratic society, it transcends the gender binaries and hierarchies that structure patriarchal institutions and cultures. An ethics of care is key to human survival and also to the realization of a global society.

It is difficult to understand how caring could be relegated to one gender or be seen as a lower contribution than making money in a democratic society where all people are considered equal—a society where caring is an essential quality that enables everyone to have a voice and be heard with respect.

In Occupational Health, the ethic of care supports the idea of removing blame from accidents and mistakes. Thus it has the potential to dramatically reduce workplace stress and foster an explosion of innovation, resonating strongly with psychological safety. The fact that most businesses have not embraced the ethic of care as part of their values is again a symptom of an individualistic culture seeking to avoid collective responsibility. An ethic of care in mature safety cultures shows up as employees actively look out for one another. The generativity research supports this by identifying being caring, helpful, and collaborative—as a key personality trait driving the generativity motive. Therefore, a generative culture is inherently a caring culture.

Generativity Comes from the Ethic of Care

The generativity framework emphasizes that generative behaviors are inherently social and relational, requiring a deep understanding of others' needs and perspectives. The meta-analysis by Doerwals, et al. (2021) indicates that organizations can cultivate generativity by providing job autonomy and challenging job demands, both of which are positively linked to the generativity motive. This offers leaders concrete levers to create a more generative and, consequently, safer environment.

Generativity is also associated with positive outcomes such as increased work motivation, occupational self-efficacy, and effective organizational commitment. This data strengthens the business case for investing in a human-centered culture. While generativity is often linked to older, experienced workers sharing knowledge, research shows there are no age limitations.

Generative Leader and Behaviors

From this core philosophy flows generative leadership and behavior: the active roles and concrete actions that express this altruistic motive. These are the leadership practices that foster growth, cultivate psychological safety, empower others, and intentionally build the strong, supportive relationships essential for maintaining the well-being of the soft infrastructure.

Generative social fields (GSFs), a concept explored by Otto Scharmer and Peter Senge, highlight the profound influence individuals have on each other within a group. A GSF creates an inclusive environment where people feel safe to express themselves without fear of exclusion, addressing a fundamental human need for inclusion.

A conscious leader can leverage generativity to cultivate a high-performance environment, becoming a generative leader. This type of leader embodies the eight beliefs of RCS, fostering growth, psychological safety, and empowerment. Generative leadership practices actively build the strong, supportive relationships essential for a sustainable and safer future.

Roles in the Organizational Immune System: Beyond the Checklist

The executives' role within this central nervous system transcends simply reviewing dashboard metrics. It involves educating management staff to cultivate employees as risk sensors to detect the intangible, constantly shifting signals of potential hazards: the quiet concerns, the workaround shortcuts, and the unwritten norms that often precede catastrophe. These subtle signals can manifest as low-grade fevers, such as increased employee turnover in a specific department, recurring minor quality issues, or a noticeable dip in cross-departmental collaboration, all signaling early distress.

An organization can be understood as a living system, with its health and resilience continuously protected by an immune system. This system's strength lies not in its physical assets but in its soft infrastructure: the vital, often intangible network of relationships, communication pathways, shared knowledge, and management practices that dictate how the organization truly functions. This framework moves beyond a mechanistic view to see the organization as a holistic, interconnected system where safety is an emergent property of its social health.

From a systems thinking perspective, these examples show how actions in one area (leader behavior) have ripple effects across the entire interconnected system. A leader's actions, like greeting employees by name, directly challenge the existing, often unspoken, assumption that "people don't care about me" or "my contributions aren't seen." When a site leader consistently acknowledges individuals by name, it's a powerful signal of recognition and value.

To better explain how the organization functions as a living system, we can analyze its operations at three social levels introduced in Chapter 5:

- *Macro-social level:* Refers to the larger elements that influence human behavior, such as culture, governments, or religions.
- *Meso level:* Relates to the level closest to individuals, including teachers, supervisors, companies, or departments.
- *Micro level:* The ground level, which refers to how individuals interact and intergroup experiences.

The Immune System in Action: A Multi-Level Analysis

Sensing and Interpreting Threats

A healthy organizational immune system relies on its ability to detect the subtle, intangible signals of potential hazards—the "low-grade fevers" that precede catastrophe. This requires actively cultivating every employee as a "risk sensor." If fear of embarrassment or reprisal silences employees, a vital sensory nerve is severed, leaving the organization blind to emerging threats.

It is not about investigating accidents after they happen but about creating a continuous organizational capability to anticipate and adapt to future surprises—a core component of resilience (Kilskar et al., 2018). This involves enabling frontline workers to notice and discuss puzzling incidents and weak signals before they escalate into full-blown failures. This transforms sensemaking from a post-mortem analytical tool into a proactive, ongoing organizational process for building resilience.

Micro Sensing: This happens at the floor level and the executive level. The executives' role within this central nervous system transcends simply reviewing dashboard metrics. It involves educating management staff to cultivate employees as risk sensors to detect the intangible, constantly shifting signals of potential hazards: the quiet concerns, the workaround shortcuts, and the unwritten norms that often precede catastrophe. These subtle signals can manifest as low-grade fevers, such as increased employee turnover in a specific department, recurring minor quality issues, or a noticeable dip in cross-departmental collaboration, all signaling early distress. This proactive cultivation of risk sensors is a key focus, emphasizing tools and methods to empower employees in this vital role.

At the floor level, this is seen in the story of Alex, a miner whose embodied cognition allowed him to sense an unmapped methane pocket through a "faint, sweet, cloying smell" and the feeling that "the air had a different taste." His personal, sensory experience acted as a vital risk sensor where technology, which showed "all green," had failed. Similarly, the employee Mike had an "uneasy feeling" about not performing a Lockout/Tagout verification, another valid individual-level signal that was tragically ignored.

Meso/Community-Level Sensing Systems: Organizations can build systems to harness this sensory network. At Opticore-Tech, the "Ride Alongs" program was a group-level strategy to immerse safety facilitators in the daily work of frontline employees, allowing them to learn the specific language of hazards and understand risks from the workers' lived experiences. The company's "I've got a question" Stop Work Authority (SWA) program empowered every employee to act as a sensor by making it psychologically safe to flag even minor uncertainties.

Collaborative Defense Mechanisms

When a threat is detected, the organizational immune system develops defenses, or "antibodies," in the form of shared understandings and practices. These defenses cannot be dictated from the top down; they must be collaboratively built through continuous conversation, trust, and shared meaning-making—the process of social construction.

Leadership's role is to foster the lymphocytes and T-cells of the organization—empowered, cross-functional teams capable of collectively interpreting threats and devising effective, real-time responses. This necessitates delegating authority, providing necessary resources, and encouraging experimentation and learning at the front lines.

Group Collaborative Defense: At Opticore-Tech, the Japanese practice of Nemawashi—holding small, informal discussions to build consensus before a formal meeting—is a clear example of a collaborative defense mechanism. It ensures shared buy-in and avoids conflict when implementing safety solutions. The entire "safety counterinsurgency" model was designed to foster trust and shared responsibility, creating a robust, organization-wide defense (case described in Chapter 11).

Individual-Level Collaborative Language: In the 3-Cs case study at a nuclear power plant, a coach helped a manager shift his team's language from "command, control, and compliance" to "Caring, Collaboration, and Co-creation". By introducing and reinforcing this new vocabulary, the team collaboratively constructed a new reality of teamwork and shared problem-solving, building a stronger local defense against error (case described in Chapter 7).

The 2005 Japanese railway crash illustrates a catastrophic failure of collaborative defense at the group level. The organizational culture used shame and embarrassment to enforce accountability, which actively destroyed the trust and psychological safety needed for collaborative sensemaking and the development of shared defenses (case described in Chapter 8).

Learning and Adapting (Immunization Memory)

A resilient living system remembers past infections to better fight future ones, a process of immunological memory. For an organization, this means engaging in continuous sensemaking to learn deeply from incidents. It involves teaching leaders how to move beyond a single root cause and sensemaking to discover how individuals made sense of the situation at the time, given their available information, pressures, and collective interpretations. Such profound learning, mirroring immunological memory, prevents the organization from repeating past illnesses and cultivates stronger, more resilient responses to new challenges.

Meso-Level Learning System: Paul O'Neill's leadership at Alcoa created a powerful organizational memory system. By demanding that accident reports be sent directly from the lowest levels within 24 hours, he created a rapid feedback loop that forced the entire organization to learn from every incident and continuously adapt its practices. This became a "keystone habit" that spread excellence throughout the company.

Macro Level: The Deepwater Horizon explosion is a stark lesson in failed organizational memory across the industry. A prior near-miss on another Transocean rig had been successfully managed, and the procedures—the "antibody"—were shared with BP executives. However, this critical information was never communicated to the Deepwater Horizon crew, demonstrating a catastrophic breakdown in the organization's ability to distribute learning and adapt (see Chapter 10).

Prioritizing Protection over Production

A critical danger is the organizational equivalent of an autoimmune disease, where the immune system attacks healthy tissue. This mirrors the monetarily-driven spiral to disaster that Bob Bea highlights, where the relentless pursuit of production imbalances and inadvertently harms the organization (2025). If safety is relegated to a burdensome, box-ticking exercise rather than an integral component of the organism's health, internal pressures can inadvertently create the conditions for catastrophe. Executives have a strategic imperative to ensure that the pursuit of growth never compromises the very health of the organization.

Meso-Level Autoimmune Disease: This pathology is evident in several major disasters. At Massey Energy's UBB mine, the CEO explicitly encouraged supervisors to focus on "running coal" over complying with safety standards. Similarly, the president of Disneyland announced a maintenance policy of running rides "to failure" to save money, a linguistic framing that prioritized finance over protection and led directly to fatalities.

Micro-Level Manifestation: The warehouse supervisor's story of a "Dumb and Tough" culture shows this at the team level. When he raised ergonomic hazards, he was told to just "make a good speech," and the prevailing attitude was that injuries were the worker's own fault for being "dumb" and should be endured without medical attention. This is a clear case of the system attacking its own members to serve production goals.

Macro-Level Pressures: The sources note that the Bangladesh Rana Plaza disaster was influenced by organizational-level market and financial pressures from the West, which created a "path of least resistance" toward prioritizing profit over worker safety. This shows how broad socioeconomic forces can induce an autoimmune response in organizations.

Relationship-Centered Safety

RCS is a framework that can help shift organizations from the mechanistic to the more adaptive model of an organic, living system. The health and resilience of a living system are not secured by top-down mandates but by the connectivity and relationships among its members. Leaders who practice RCS instinctively grasp this. This perspective allows them to move beyond individual accountability to foster the collective intelligence and proactive defenses necessary for the entire system to thrive.

RCS is a philosophy detailed in The Relationship Factor in Safety Leadership (Carrillo, 2019). The framework is based on relationship psychology, a branch of complexity management that looks at organizational outcomes because of social interactions and relationships (Stacey, 2003). It contains many stories of leaders who used a relationship-centered style to achieve excellence in safety, quality, and production. Such leaders naturally view the organization as a living system, a model introduced in this book to shift from an individual to a collective approach to accident prevention. This chapter makes the principles accessible to leaders who need to learn and apply them.

Generative leaders act from their beliefs about human nature. Beliefs create thoughts and thought informs action. A belief is not a passive thing like a value. Values can be espoused but not practiced. Beliefs are so deep that you sometimes don't even know they are influencing your decisions. It is not a matter of having faith that something is true; it is knowing that it is true. Our understanding of the nature of relationships and how it affects safety performance could be the defining factor between success and failure (Housel, 2023). The eight beliefs of RCS reflect what successful leaders believe they know about people. These concepts are philosophical but also based on research in cognitive psychology as captured by Argyris' Ladder of Inference (1982).

Below are two strategies demonstrated in this book: RCS and the power of small incremental changes.

Relationship-Centered Leadership and Small, Incremental Changes

In a living system, significant change often arises not from massive overhauls but through small, consistent, positive interactions that create ripple effects. These small nudges can rewire the social system and build trust more effectively than large, top-down mandates.

At a Rio Tinto mine, the General Manager fostered a culture of care with a simple directive: he instructed his managers to greet every employee by name. This small but consistent act transformed a low-morale environment into one of high trust and engagement. He also united the site with a generative vision of defending Australian manufacturing, giving meaning to the work beyond individual paychecks.

At the nuclear power plant, a manager was coached to start by simply using team members' names in meetings. This small shift in interaction was immediately noticed and appreciated by his team, laying the groundwork for deeper trust and collaboration (see Chapter 7).

Group-Level Linguistic Nudges: At Opticore-Tech, renaming "safety technicians" to "safety facilitators" and "audits" to "Ride Alongs" were small linguistic changes that signaled a profound philosophical shift away from enforcement and toward collaboration and support. These actions perfectly illustrate how small, consistent changes by leaders can lead to new patterns of interaction and behavior (see Chapter 11).

The Power of Small Incremental Changes in the Social System

The examples of successful change using this strategy came about as a result of leaders making small consistent positive changes in their interactions with organizational members. In the Rio Tinto Mine Case Study, the site manager instructed his direct reports to say hello to every employee they passed and use their name. Similarly, Opticore changed the language from safety tech to facilitator and Ride Alongs instead of audits and dramatically improved the relationships and performance.

The case examples (Opticore-Tech and the 3-Cs) where small changes in language or repeated interactions improved communication through relationships perfectly illustrate that subtle, indirect suggestions and alterations to the environment can encourage desired behaviors more effectively than strict mandates. These are incremental changes—small, manageable adjustments that are easier to implement successfully and foster sustainable progress. Instead of a grand, pre-planned transformation, these small, consistent actions by leaders lead to new patterns of interaction and behavior, ultimately raising trust levels and open communication.

Complexity theory explains how these small changes can achieve big results. Organizations are dynamic systems where small fluctuations can lead to significant, often unforeseen, changes (Stacey, 2003). A leader's consistent, positive actions in the field of social interactions creates a shift in the social system's dynamics, allowing relationships to self-organize into a more coherent and productive state. However, the small incremental actions will have no effect unless they are followed up with responsiveness to concerns, aligned resource allocation, consistent leadership, and robust systems to manifest as real-world results and avoid cynicism.

Managing Uncertainty

Complexity theory explicitly asserts that organizational behavior is inherently unpredictable and non-linear. The significant conclusion drawn from this is that instead of attempting to suppress or fight this inherent unpredictability, the soft infrastructure model actively incorporates it into its design. An example would be a scenario where deviating from a procedure is seen as a natural human inclination to innovate, improve, or to feel safer. Instead of labeling it as undesirable, it could be embraced as a normal process. In this updated process, individuals would openly discuss the changes they've made, receive feedback on any potential negative consequences, and be recognized for their procedural enhancements.

It is not about tolerating disorder, it's about shifting management's role from direct control and rigid goal-setting to facilitating the conditions under which good ideas and adaptive practices can be recognized, experimented with, and integrated. Instead of controlling, the leader recognizes and leverages people's natural inclinations.

The traditional, control-oriented leader designs rigid structures and enforces compliance to control uncertainty. The Opticore-Tech "safety counterinsurgency" model directly demonstrates systematically shifting its approach from control and enforcement to facilitation and relationship-building, directly embodying the principles of RCS.

Shifting from Control to Facilitation: Opticore-Tech deliberately renamed its "safety technician" to safety facilitators. This change in language signaled a profound shift away from a perception of safety personnel as enforcers and toward a culture of collaboration, support, and shared problem-solving.

Understanding Work-as-Done: The facilitators conducted Ride Alongs with frontline workers. The explicit purpose was not to audit or evaluate performance but to immerse themselves in the daily work, learn the workers' language, and understand challenges and hazards from their lived experience. This is a direct application of understanding and leveraging natural inclinations rather than enforcing rigid rules from a distance.

Psychological Safety: The transformation of the SWA-policy is a prime example of how to deal with fear and uncertainty. Instead of an intimidating, bureaucratic process, Darren Brooks of Opticore-Tech reframed SWA into a simple, natural act of team solidarity: asking a question. The phrase "Hold on, I got a question" was normalized, removing fear and encouraging employees to voice uncertainty about even small things. This empowered workers to innovate and feel safer by making it easy to pause and discuss deviations or concerns, which were then addressed collaboratively. This builds a formal process for managing, rather than suppressing, unpredictability.

Leadership Competencies—Primary Responsibility Is to Monitor and Nurture the Health of Living System

Leading within this living system requires a specific set of beliefs about human nature and related competencies. The health of the organization's "immune system" depends on a leader's ability to demonstrate commitment not through technical expertise but through the very social and problem-solving skills that allow them to effectively interpret signals from the "risk sensors" and foster collaborative defenses.

The metaphor of the organization as a living system guides executives in recognizing and investing in the organization's "social architecture." To translate this dynamic, relational concept into a resilient reality, leaders need a strategic blueprint that aligns their core beliefs, organizational design, and daily actions. This framework is achieved by establishing a path that begins with foundational assumptions about human nature, followed by an organizational structure, and finally, the necessary leadership behaviors.

The Strategic Blueprint: Structuring Safety from Belief to Behavior

Building a resilient organization requires more than simply implementing procedures; it demands a philosophical shift, an organizational design change, and a daily commitment from leadership. The RCS framework provides the strategic blueprint for this shift, translating foundational organizational theories into an actionable structure for safety.

Safety excellence is realized by aligning three core elements hierarchically: a philosophical belief in the workforce (Theory Y), an organizational structure that empowers them (Socio-Technical Systems (STS)), and the necessary daily behaviors that activate the system (The Eight RCS Beliefs).

1. The Foundational Belief: Theory Y (The Philosophical Why)

The philosophical starting point for a human-centric safety system lies in Theory Y, proposed by Douglas McGregor. This belief system asserts that workers are not inherently lazy or irresponsible (the old Theory X belief). Instead, Theory Y posits that individuals are capable of self-direction and self-control, and will actively seek responsibility if they believe in the goals of the enterprise.

For leadership, adopting Theory Y is the crucial philosophical shift required to move beyond a compliance-only mindset. This positive assumption directly supports RCS Belief #4: People are able and willing to contribute to the success of the enterprise. When leaders operate from this belief, they foster the sense of inclusion and belonging necessary for collective contribution.

2. The Structural Design: STS (The Organizational How)

The Theory Y philosophical belief must be supported by an organizational design that structurally enables it. This is the role of the STS model.

The STS model is founded on the principle that an organization is composed of interdependent technical systems (equipment, processes) and social systems (people, relationships). The goal of STS is to achieve the "ideal balance" between people and technology. It dictates that work must be designed to reflect the competence of the people doing the work, creating self-managing teams that are structurally empowered to control the technical process.

This design component is crucial for moving safety from a top-down mandate to an emergent property of the organization. As one pioneering researcher noted, the maintenance of safe operation is an "interactive, dynamic and communicative act" that is vulnerable to disruption by interventions that ignore the processes by which safety is constructed. STS provides the structure that acknowledges the human element and facilitates this construction.

The Activation Mechanism: The Eight RCS Leader Beliefs (The Daily Action)

Theory Y and STS are theoretical models; the Eight RCS Beliefs are the practical, daily behaviors (the essential "glue") that leaders must consistently exhibit to operationalize the theory and make the self-managing structure successful.

These beliefs ensure the system works by reinforcing trust and psychological safety. For instance, Belief #1, True communication takes place in the presence of relationship and trust, guides leaders to prioritize relationship-building activities like "Ride Alongs". These actions create the environment where workers feel safe to report errors, exert control, and engage in the continuous sensemaking necessary to prevent accidents. The success of the Opticore-Tech model in cutting accidents by 50% demonstrates the power of implementing these beliefs to transform structure (e.g., reframing SWA to "I've got a question") and thereby activate collective responsibility.

Motivation for Leaders: Justification and Generativity

This structural framework motivates leaders because it provides both the rational justification and the deep moral motive for change:

Rational Justification: Theory Y provides the conviction that distributing power to the front line through STS is the rational path to superior performance. Empirical evidence supports this, showing that a senior manager's commitment to safety is demonstrated primarily by their problem-solving and social competence (specifically perceiving others), not their technical safety knowledge. By focusing on social systems and relationships, leaders are investing in the measurable skills proven to drive safety commitment (Fruhen et al., 2014).

Moral Motive (Generativity): While Theory Y is the philosophical justification, Generativity provides the powerful, intrinsic motive for leaders to sustain this effort. Generativity is the desire to support and guide younger people and to benefit "future generations." This desire to leave a positive legacy forms the ethic of care, which is reflected in RCS Belief #8: Relationships influence emotions, feelings, and beliefs, which influence decisions. This moral commitment ensures the leader prioritizes the well-being of people over profit, a failure that often precedes catastrophe (Doerwald, et. al., 2021).

Conclusion

This chapter has provided a strategic framework for shifting OHS from a reactive cost center to a foundational element of organizational health and resilience. Through the metaphor of the organization as a living system, we have moved beyond a mechanistic view to one that recognizes that "at the core of organizational longevity is the interaction of social, emotional and cooperative whole human beings in various kinds of personal relationships to each other" (Schein & Schein, 2018).

Ultimately, safety is an emergent property of the system's health. Nurturing this health is a strategic imperative that requires leaders to move from a top-down, command-and-control mindset to one that trusts people at the job site to make informed decisions. By educating and enabling every individual to fulfill their roles and communicate effectively, leaders can move beyond checklists and create a high-performing organization that is both deeply human and highly resilient.

Executive Takeaways

If you adopt the "Living System" metaphor what changes in your leadership mindset? What does it mean to shift from a mechanical view to an organic one? Recognize that safety is not a department or a checklist but an emergent property of the organization's health and the network of relationships within it.

Enable Decentralized Decision-Making. Move authority to where the work is done. Educate and trust people on the front lines to make informed decisions. A clear protocol for what to do when you don't have an answer is essential.

Invest in Soft Infrastructure. Reframe relationships, trust, and open communication as core strategic assets. Actively monitor the health of these vital systems to uncover hidden risks, as seen at AES.

Embrace an Ethic of Care. Recognize that a safety culture rooted in empathy and care is a strategic imperative. By valuing compassion and genuine concern for well-being in decision-making, you can build a more resilient and sustainable enterprise.

References

Argyris, C. (1982). *Reasoning, Learning, and Action: Individual and Organizational*. Jossey-Bass.

Bakke, D. W. (2005). *Joy at Work: A Revolutionary Approach to Fun on the Job*. PVG.

Bea, R. (2025). *Robert Bea: Dr. Disaster*. UC. Berkeley Engineering. https://engineering.berkeley.edu/robert-bea/

Carrillo, R. A. (2019). *The Relationship Factor in Safety Leadership*. Routledge.

Doerwald, F., Zacher, H., Van Yperen, N. W., & Sheibe, S. (2021). Generativity at work: A meta analysis. *Journal of Vocational Behavior*, 125: 103521.

Erikson, E. H. (1950). *Childhood and Society*. New York, NY: W. W. Norton.

Erikson, E. H. (1982). The Life Cycle Completed. New York, NY: W. W. Norton & Company.

Fruhen, L., Mearns, K., Flin, R., & Kirwan, B. (2014). Skills, knowledge and senior managers' demonstrations of safety commitment. *Safety Science*,69: 29–36.

Gilligan, C. (2011a). *Join the Resistance*. Polity Press.

Gilligan, C. (2011b). *Ethics of care*. https://ethicsofcare.org/carol-gilligan/#:~:text=An%20ethics%20of%20care%20directs,rather%20than%20deduc-

Katz, D., & Kahn, R. (1978). *The Social Psychology of Organizations*. Wiley.

Manz, C. C., & Sims, H. P. Jr. (2001). *The New Superleadership: Leading Others to Lead Themselves*. Berrett-Koehler.

McShane, S., & VonGlinow, M. (2021). *M: Organizational Behavior*. McGraw Hill.

Pawar. M. (2025). Organizations as living systems: Lessons from human physiology. *Systems Thinker*. https://thesystemsthinker.com/organizations-as-living-systems-lessons-from-human-physiology/

Schein, E., & Schein, P. (2018). *Humble Leadership*. Berrett-Koehler.

Stacey, R. (2003). *Complex Responsive Processes in Organizations*. Routledge.

Ullberg, S. B. & Warner, J. (2016). The relevance of soft infrastructure in disaster management and risk reduction. *UN iLibrary. UN Chronicle*, 53(3): 23–26.

Ullberg, T., Zia, E., Petersson, J., & Norrving, B. (2016). Doctor's follow-up after stroke in the south of Sweden: An observational study from the Swedish stroke register (Riksstroke). *European Stroke Journal*, 1(2): 114–121. https://doi.org/10.1177/2396987316650597

Wetlaufer, S. (1999, January). Organizing for empowerment: An interview with AES's Roger Sant and Dennis Bakke. *HBR*. Download 8/27/2018.https://hbr.org/1999/01/organizing-for-empowerment-an-interview-with-aess-roger-sant-and-dennis-bakke

Wheatley, M. J. (2006). *Leadership and the New Science*. Barrett-Kohler.

Wheatley, M. J. (1997). Goodbye command and control. *Leader to Leader*, (5): 21–28.

10 The Social Anatomy of Deepwater Horizon and Its Legacy

Introduction

As we engage in understanding the Deepwater Horizon (DWH) disaster, there is a prevailing sense of déjà vu—we conduct investigations, submit reports, yet substantive change remains elusive. The Chemical Safety Board released several volumes of their report on DWH, the first in 2014, identified systemic shortcomings in culture, leadership, risk management, and corporate social responsibility (CSB, 2014). Two years later the CSB released volume 3, which found that offshore risk management and regulatory oversight remained inadequate despite the 2010 explosion (CSB, 2016). The mantra of unacceptable behavior and inadequate risk management or elements of blame levied against people do not result in meaningful or sustainable change, yet they typify our response to similar incidents (Douglass, 1992).

Emerging from the seemingly heartfelt admonishment by regulators, legislators, and the public that "we must end this" the problem remains pretty much unchanged. Oil production has returned to pre-disaster levels, shifting the emphasis from change to justifying the risk for economic stability. This reinforces the notion that risk is socially constructed and accepted; in this case, economies dependent on deep-sea drilling socially constructed a reality of "acceptable risk."

Is there value in continued reflection? This book has consistently argued against the myth that individual responsibility is the key to safety. Now we examine a more insidious cause: the socially constructed blindness of the executive 'VIP visit.' By analyzing the tragedy through the lens of ritual, social hierarchy, and 'survival mode,' we unmask how an obsession with personal injury rates blinded an entire organization to a looming catastrophe—and why expertise becomes inaccessible when trust collapses.

The Social Construction of the VIP Visit

The National Commission (2011) report identified the central cause of the accident as a failure of management, calling for "unwavering commitment to safety at the top of an organization: the CEO and board of directors must create the culture and establish the conditions under which everyone in a company shares responsibility

DOI: 10.4324/9781003668770-10

for maintaining a relentless focus on preventing accidents (218)." Research confirms that this commitment is best demonstrated through social/relational interaction coupled with problem-solving skills—actively seeking cognitively diverse information and perceiving others' emotions—rather than only technical knowledge (Fruhen et al., 2014; Page, 2011).

A logical, yet insufficient, conclusion would be to prescribe more standard leadership development for executives to strengthen their ability to communicate commitment. However, based on psychological and neuroscientific frameworks, the executives onboard that day were likely genuinely unaware of the critical tensions regarding the well's stability or the system variability that was soon to emerge. This unawareness was not random; it was a socially and cognitively constructed blindness—a normalization of risk (Pupulidy, 2015).

From a social psychology perspective, the rig environment on that day was structured to obscure reality. The "VIP visit" is a powerful social frame that alters normal behavior. In Goffman's terms, the crew and visitors engaged in a "performance." the crew stayed on the "front stage" where compliant and expected behaviors were presented. The intense arguments over negative pressure tests occurred "back stage," hidden to maintain the appearance of competence.

The executives, seeing the calm demeanor of their peers, likely concluded, "No one else looks worried, so there must be nothing to worry about." They were primed to celebrate a seven-year safety record, and their brains, hyper-focused on positive stimuli, failed to perceive contradictory signals—a classic case of inattentional blindness (the failure to notice a fully visible but unexpected object or event because attention is focused on another task or event). Every smiling handshake confirmed their pre-existing belief in excellence, while ambiguous signals of stress were dismissed as standard rig noise (confirmation bias).

DWH Stands as a Definitive Example of a Reliance on Individual Responsibility

It was not a failure of individuals to take responsibility; it was a complete systemic social collapse that rendered individual responsibility futile. The record proves that individuals *did* speak up. BP wells team leader John Guide explicitly warned David Sims, a BP manager that the crew was at their "wits' end" and flying by the seat of their pants (CSB, 2014). Yet, these warnings were swallowed by a social system prioritized to cut costs. A blatant example was when a three-man Schlumberger team was sent home before they completed a test to measure the strength of the cement bond log in the well. BP sent them home the day before the explosion to save a $128,000 fee (Averill et al., 2022).

Individual responsibility was used as a legal shield when the inevitable catastrophe occurred. The system did not examine its own social collapse; instead, it weaponized individualism to absolve itself in an act of "legal theater." The system held and prosecuted two frontline individuals, Robert Kaluza and Don Vidrine. While BP pleaded guilty to 14 criminal charges and paid $65 Billion USD in fines, at the end of the day none of the corporate leaders who were responsible for the design and overview of the safety conditions were put on trial.

The Deepwater Horizon tragedy stands as the definitive case for redefining the executive leader's role as a truth seeker. Former Transocean leader, Thomas Hinterseer reflecting on his experience before, and after the disaster offered this insight: "safety is **the presence of leadership behaviours that relentlessly surface reality.** It is built through humility, proximity to the frontline, and cultures where speaking up is an act of strength, not defiance. (Hinterseer, 2026b)."

The "Zero" Paradox: How Success Blinded the Organization

The executives onboard the DWH that day were not negligent in their own minds; they were blinded by a specific social construct known as the "Zero Paradox." As noted by Dr. Robert Long, an obsession with "Zero Harm" and low personal injury rates can create a "culture of hiding" and "risk blindness" (Long, 2012). BP's intense focus on minor personal injuries—slips, trips, and falls—created an illusion of safety. Because their injury numbers were low, the organization felt safe, effectively silencing the "weak signals" of major systemic risk (Long, 2012).

This helps explain the cognitive disconnect of the VIPs during their visit. The executives were celebrating seven years without a lost-time incident. In their social reality, "safety" was defined by the absence of personal injury. Long argues that this "Zero discourse" blinds the organization to catastrophic risk because it creates a social reality where "bad news" is suppressed to maintain the illusion of Zero (Long, 2012).

The tragedy was compounded by a stark irony: the VIPs on the rig that day were not merely administrative "suits" there for a photo opportunity; they were the organization's top technical experts. Patrick O'Bryan, the Vice President of Drilling and Completions, was physically present on the rig floor. These men possessed the exact technical competence required to recognize the well-stability issues and authorize the necessary delays (Hopkins, 2011). Yet, in the face of mounting danger, no crew member approached them for help.

Why did the crew and managers remain silent in the presence of their best resource? Perhaps because the prior warnings from the rig leadership were met with inaction, a fatal message was broadcast to the workforce: You are responsible for safety—you are on your own. Consequently, the crew viewed the VIPs not as resources to be consulted but as authority figures to be placated, proving that when trust collapses, expertise becomes inaccessible even when it is standing right next to you.

The crew, understanding this cultural mandate, engaged in a performance of safety. To speak up about the well instability would not only have ruined the celebration but would have shattered the "Zero" narrative the executives relied upon for their sense of security. Thus, the organization lost what Dr. Long terms "risk imagination"—the ability to foresee a blowout—because their language, rituals, and rewards were entirely focused on the maintenance of zero personal injuries.

The Absence of Embodied Cognition

While humans possess mirror neurons that allow us to "catch" others' emotions, social power dynamics can override these biological systems (Hatfield et al.,1993).

Even if they had sensed tension, seasoned executives often normalize high-stress environments as business as usual. Research indicates that high power can neurologically dampen empathy, a form of embodied cognition. The crew engaged in emotional labor, replacing accurate data (fear, stress) with false data (calmness, smiles). The executives were looking at a dashboard of human faces that all read "Normal," even though the internal systems were flashing "Critical Failure." Combined with the crew's intense emotional labor—acting calm while terrified—the noise of the celebration effectively drowned out the signal of impending disaster.

The failure to perceive these signals was not merely a cognitive oversight; it was reinforced by a powerful hierarchy. Hopkin quotes from one of the interviews to explain why the executives did not interact with other people on the rig,

> You are managing a group of professionals who have very clear responsibilities […] The implication here is that to question what they are doing is to doubt their professionalism […] any observation would go down through the chain of command […] no direct contact with crew members.
>
> (2:8) 2012:8

While the executives later claimed they saw only "smiling faces," the written record tells a different story. Just three days before the disaster, visiting executive David Sims received an email from BP wells team leader John Guide, explicitly stating that the rig crew had finally "come to their wits' end" and that the operation was "flying by the seat of our pants" (Hopkins, 2012).

Despite this direct plea indicating the system was in distress, the VIP visit proceeded as a celebratory ritual rather than an inquiry. By trying not to "undermine authority" or disrupt the visit, the VIPs were fortifying the silence. Their lack of inquiry was a loud statement that "saving face" was the highest priority. This professional distance effectively masks a system's internal collapse with a veneer of institutional rectitude—often mistaken for professionalism.

The oil industry presents itself as secular-rational, driven by data, engineering, and objective rules rather than tradition or religion. However, the VIP visit itself operated as a traditional ritual, compelling strict conformity. Within hierarchical value systems, deference to authority is supreme. The executives were essentially the high priests of the corporation, arriving to bestow a blessing upon the rig for its safety record.

To reveal the inconsistent well test results during this ritual would have been an act of heresy, violating the expected deference. Not to mention that the "hero" mentality of the oil and gas industry was "Yes we will get the job done." Admitting weakness or reducing speed, was against the mentality (Hinterseer, 2026a). Furthermore, the executives might have been hesitant to appear critical or disempowering by asking questions about what was going on the rig. This created a dilemma where social pressure superseded created blinders.

The "myth of individualism" suggests that if executives just cared more, they would have seen the danger. Social science suggests otherwise: the hierarchical chasm made authentic sensemaking impossible. Although the VIPs on board were experts in the areas where the crew was struggling, no one approached them for help. Stepping outside established roles during such a high-stakes ritual would have risked severe loss of face for all parties involved.

However, engaging in group sensemaking or even questioning the workforce would have identified the vulnerable nature of humans working in high-stress, high-outcome scenarios and may have allowed for a collective response in time to prevent catastrophe. However, the executives were most likely not capable of taking on a helping role due to the high power neurological states that can actively dampen empathy and the ability to perceive the need to help for many high status individuals (Hogeveen, et al., 2014).

In summary, executives failed to grasp the critical issues being faced by the crew of Transocean. Mistaking a workforce in survival mode for one operating transparently poor social perception prevented them from interpreting the subtle emotional cues around them. Or they may have mistaken them as anxiety in the presence of VIPs, rather than concerns about their safety. The next section takes a closer look at the whole system and the socio political forces that led to the event.

The next section takes a closer look at the whole system and the socio political forces surrounding the industry.

Flawed Decision-Making and Planned Ambiguity

The crew's inability to see the full picture led to disastrous decisions, influenced heavily by organizational pressures, culture, and a fear of losing face. Dissenting voices did exist but were effectively silenced. Jimmy Harrell was the offshore installation manager and the top drilling supervisor for rig owner Transocean. He testified that he had expressed concerns about the drilling plan, which he believed was rushed by BP, but he ultimately conceded (Hammer, 2021). His grim comment, "Guess that's what we have those pinchers for" (referencing the pinchers on the blowout preventer) signals a breakdown of shared decision-making (Goodman, 2013). It highlights a system where individuals, despite their expertise, felt powerless to stop the momentum of production pressure that dominated the culture.

The BP Vice President of Drilling, Patrick O'Bryan, testified that the corporate safety system (operations management system (OMS)) was, at best, "equivocal" in its application. This ambiguity in procedures was not accidental but served as a potential legal shield. By maintaining that the OMS did not directly govern contractor rigs, BP could argue that the contractor (Transocean) and its safety system were primarily at fault in the event of an incident (Goodman, 2013).

The fact that the rote, but ambiguous, procedures were meant to be handled by the people on board places the full burden of navigating systemic uncertainty onto the frontline crew members, like Vidrine, who proceeded with the fatal displacement operation despite a clear and unambiguous verbal warning from an onshore engineer that the integrity test results were "physically impossible for a secure well."

In summary, the defense-in-depth system failed because its barriers were not independent; they were all compromised by the same flawed, socially constructed assumption that "all was well" (Hopkins, 2023). They also failed because the risk was viewed simplistically as energy to be contained and that the defenses in depth, coupled with hierarchy of controls and prevention through design, had addressed the issues. "Risk management strategies indicated the system was safe enough (Pupulidy, et al., n.d.)." This tragedy was the direct result of an individualistic culture. In such a culture, executives would not have been expected to engage in providing guidance or support so that they could be approached. It is unlikely that events like this could be handled collectively if it were not part of executive education or conversations.

The cry for accountability was heard around the world. A major loss were the 11 individuals who lost their lives and the 17 injured; however, watching the footage day after day of the damage to the environment and harm done to animals was also heartbreaking. Finding what or who to blame became critical despite a lack of information. If the VIP visit on the rig that day was a social performance, the prosecution was the next act of a play that valued holding people accountable over understanding and learning.

The systemic, organizational failures were narrowly translated by the Department of Justice prosecution into the criminal gross negligence of two individuals, Don Vidrine and Robert Kaluza, who were physically on the rig. The prosecution of only two people at the site strongly reinforces this book's argument that individualism creates the "accountable individual" to protect the underlying power structure from scrutiny. Eventually those charges were dropped for lack of evidence (Archives, 2012). This was not a simple failure of Transocean, or BP, it was a failure of the entire offshore oil production system.

The prosecution of the two on-site workers was characterized as "legal theater—a 'scapegoat' mechanism", which successfully diverted public attention from the senior executives who were responsible for designing the flawed systems, cost-cutting measures, and ambiguous safety protocols that were the network of both influences and causes leading to the disaster. The result was that nothing needed to change. And eventually everything went back to "normal" (Kaluza & Karinch, 2020).

A Culture of Fragmented Awareness

Did individualism play a major role that led to the well blowout that fateful day in 2010? I am going to say yes. An individualistic culture structurally isolates people, assumes individuals should just do their own jobs, and does not spend time developing relationship capacity. This ultimately makes it impossible to transmit the weak signals vital for true situational awareness. Among these signals are emotions; as we know, people in that industry are pressured to not show weakness.

In short, the rig had a culture of fragmented awareness that shattered th organization's intelligence-gathering sources. I am calling it a culture of fragmented awareness because siloed information was baked into the way the rig conducted business. This was demonstrated by the fact that Transocean had experienced a similar event with a well that they caught in time to avoid a disaster, but that information never reached DWH. The cries for help were not answered via email or in person. The people on the shore were focused on commercial schedules, leaving the rig crew to figure out what the pressure readings meant leading to the incorrect theory of a "bladder effect." This is what happens when you leave people alone: they want to do their job, so they conduct sense-making sessions on their own. Without the expertise of the corporate structure, they were abandoned—basically left to die or, at minimum, be severely traumatized.

The Inglehart-Welzel Cultural Map

The Inglehart-Welzel Cultural Map (World Values Survey, 2023) dives deeper, offering further insight into the executive blind spots that contributed to the event. It highlights a clash between the corporate values of the visitors and the actual values governing the rig. The map uses the term *Self-expression* values to describe beliefs that emphasize individual agency, identity, autonomy, and the freedom to openly communicate one's thoughts, feelings, and personality. *Survival values* are characteristic of societies where people focus on basic needs like food and shelter; this often correlates with lower trust and tolerance, and a more ethnocentric outlook.

- **Survival Values (The Crew):** The rig was 43 days behind schedule and millions over budget. The crew operated in extreme "survival mode," driven by economic insecurity. In this mode, priorities shift to rigid conformity and protecting one's job. You do not raise complex, expensive problems when operating in survival mode. The crew survives within a hidden economy. They have limited political capital to spend. If you spend it reporting a concern that turns out to be false or that gets no backing, you might be the first to be let go or moved to a job no one else wants. Silence is the cheapest option, while the truth is an expensive luxury.
- **Self-Expression Values (The Executive Experience):** In their high-status roles, the visiting executives likely operated under assumptions of Self-Expression Values, believing that an "empowered" crew would speak up if something were wrong. This was likely a sincere belief, as executives and managers at that level spend a great deal of time talking about the importance of psychological safety. Unfortunately, neither the executives from BP nor Transocean conducted meaningful face-to-face conversations on the rig. They mistook the silence of survival the silence of well-being (things are going well).
- **Failure to Learn:** The executives failed to sense or learn about the critical issues being faced by the crew. They did not recognize a workforce in survival mode—one that was hiding information. Their poor social perception skills prevented them from interpreting subtle emotional cues as the weak signals of potential failure they truly were.

Review of the DWH disaster exposes that the deficiencies leading to the disaster were not only technical but failures in non-technical skills (NTS) such as sensemaking, communication, and learning. These core elements of NTS can be viewed as organizational and individual capacities which can be cultivated by leadership (Woods in Holnagel et al., 2006).

Most post incident investigations pointed to individual failures and inadequacies (e.g. the drill floor crew lacked "accurate situation awareness" (IOGP, 2014; Roberts et al., 2015). These labels seem accurate to many of us who have been conditioned by traditional investigative processes. The use of this language keeps us rooted in fixing holes, and often only results in the admonishment of personnel for things judged to be failures. A sensemaking lens would best foster learning and the development of NTS needed for prevention.

One of Roberts findings was a "failure to learn." A successful blowout prevention procedure was developed after a prior near-miss on another Transocean rig—but it was never communicated to the crew on the DWH (Brown, 2010). A sensemaking approach to incident review would explore the cultural influences that results in communication being siloed in organizations and the subsequent cost to organizational awareness. Studies of this kind expose system-wide vulnerabilities which are often masked until there is a catastrophic outcome. Sensemaking at an organizational level requires the sharing of information (Pupulidy, 2015).

Lacking engineering support from the shore, the crew engaged in incomplete sensemaking to explain the danger. They latched onto the "Bladder Effect" to explain anomalous pressure readings. The term was suggested by Jason Anderson, a Transocean tool pusher, to explain the anomalous pressure readings during the negative pressure test. The mental image that may have inspired the term relates to the physical components and circumstances of the well at that moment.

A key piece of the BOP stack is the annular preventer, which uses a large, flexible rubber element that inflates (like a bladder) to seal around the drill pipe or the wellbore itself. The crew member hypothesized that heavy drilling mud trapped above this seal was compressing the rubber element, which in turn transmitted pressure to the drill pipe, creating a false reading. The "bladder effect," while sounding vaguely technical, was intuitive, though fundamentally incorrect.

On the other hand, BP engineers increased the confusion and pressure by making decisions that increased the risk of cement failure, viewing these as manageable *commercial* risks rather than catastrophic *safety* risks. This unfortunate misjudgment created a false sense of security, overlooking how a business decision could quickly escalate into a life-threatening hazard.

Looking for Insights and a Way Forward

DWH illustrates how the lack of social perception skills contributed to the failure of sensemaking. Therefore, training for senior managers should prioritize developing problem-solving abilities and social perception.

Where do we look for guidance? We see these concepts put into powerful practice by Human Organization Performance (HOP). Pupulidy and Vesel (2023), describes Human and Organizational Performance (also known as Human Organization Potential) as a philosophy for integrating safety and operational excellence by treating human error as a symptom of systemic issues, not a root cause. Understanding error requires a proactive, learning-focused culture coupled with system-focused improvements, a learning-based approach to accidents, and specific metrics to track progress. The goal of HOP is prevention through learning.

The intention here is not to delve into how to implement HOP, but to provide another framework that utilizes many of the concepts covered in this book.

Leading up to the DWH event "experts on the front line" were overridden by shore-based managers focused on commercial schedules. HOP would consider this to be an example of goal-conflicts and a network of influences that prevented the social process of questioning and listening (learning). Reuben McDaniel (2007) and McDaniel et al. (2003) posited that complex adaptive systems require interventions based on sensemaking, learning, and improvisation, which doesn't guarantee perfection, but it dramatically increases the reliability of an organization's understanding of anomalies and surprises. The DWH explosion serves as a costly, yet invaluable, lesson: major industrial accidents are rarely just about technical failures. They are profound demonstrations of how a lack of understanding of social dynamics—particularly failures in sensemaking, social construction, and NTS—can lead to catastrophe.

Conclusion

The DWH tragedy exposed the fragility of psychological safety and stated corporate values when placed under pressure. It stands as a definitive case study of a catastrophe driven not merely by process failures but by a critical absence of NTS at the executive level.

The central tension of the disaster lies in defining the nature of the executive failure: Was it individual negligence or systemic inevitability? The situation was complex; the VIP visit was designed to be a social ritual choreographed to deliver good news, effectively masking the operational reality. Under these norms, executives were not conditioned to detect the crew's underlying anxiety; indeed, only a leader with exceptional emotional attunement—an outlier—would have taken note of those faint signals.

However, their failure to sense the tension was not purely accidental; it was a socially and cognitively constructed blindness. This may be why the research shows that in high-hazard environments, a leader's true commitment to safety is judged by their empathy and problem-solving acumen.

The core failure, therefore, was a deficit in social perception—the ability to "read the room" and decode the emotions and intentions of others. The executives failed because they were physically present but socially blind. They missed the low-grade signals of escalating risk because their own cognitive frameworks—and the organization's culture—were designed to suppress them.

Four Takeaways for Today's Executives

1 What are your corporate stated values. Do they differ from practiced values? An official corporate culture of "Self-Expression" (openness, empowerment) is fragile. Under pressure, the practiced culture—in this case, a "survival mode"

of fear and deference—will dominate. The crew wasn't living company values; they were trying to survive the VIP visit. A leader must constantly ask, "What are the actual pressures my team is facing right now, and which culture is winning?"

2 What are the executive leadership rituals in your company?
The DWH VIP visit was a ceremony to celebrate a safety record, not a process to audit it. This ritual created intense social pressure to provide a "good show" and made sharing bad news impossible. As a leader, you must be able to distinguish between a performance and an authentic display of conditions. While it would be difficult to deviate from the celebratory mood, leaders who notice nuances are better equipped to ask open questions, not to criticize, but to allow people an opportunity for true expression.
Attaining this ability requires self-awareness and self-development. There is no checklist to achieve this level of transparency.

3 Social perception (the Sensitivity to emotional states, reading body language) is a difficult skill to learn; it is not a soft skill.
The greatest weakness of the visiting executives was their lack of social perception. They were physically present but socially blind. They couldn't "read the room" and mistook a workforce paralyzed by fear for one that was calm and in control. This is the critical, non-technical leadership competency that was missing. It is a vital safety skill that must be integrated into all management development.

4 Executives must officially abandon the ideology of individual responsibility as the primary safety control, recognizing it as a flawed myth that obscures organizational problems and protects the power structure. Accountability must instead be framed around leadership's responsibility for creating the social relations of work that enable critical information to flow.

References

Archives US Department of Justice. (2012). BP Exploration and Production Inc. Agrees to Plead Guilty to Felony Manslaughter, Environmental Crimes and Obstruction of Congress Surrounding Deepwater Horizon Incident. www.justice.gov/archives/opa/pr/bp-exploration-and-production-inc-agrees-plead-guilty-felony-manslaughter-environmental

Averill, L., Durkin, B., Chu, M., Ougradar, U., & Reeves, A. (2022, June). *UK Loss Prevention Bulletin*. 285. University of Bradford.

Bea, R. (2011). *Final Report on the Investigation of the Macondo Well Blowout*. Deep horizon study group. www.dco.uscg.mil/Portals/9/OCSNCOE/Casualty-Information/DWH-Macondo/DHSG/DHSG-DWH-Investigation-Report.pdf?ver=I-lV-nwDpczeZsPk6JokoQ%3D%3D

Brown, R. (2010). After another close call Transocean changed the rules. *The New York Times*. www.nytimes.com/2010/08/17/us/17transocean.html#:~:text=Aug.,%2C%20the%20emergency%20equipment%20functioned.%E2%80%9D

Chief Counsel's Report (2011). *Macondo: The Gulf Oil Disaster*. www.govinfo.gov/app/details/GOVPUB-PR-PURL-gpo4390.

CSB, (2014). *Macondo Investigation Report*, Volume 1, Report No. 2010-10-I-OS, U.S. Chemical Safety and Hazard Investigation Board (CSB).

CSB, (2016). *Macondo Investigation Report*, Volume 3. www.csb.gov/the-us-chemical-safety-boards-investigation-into-the-macondo-disaster-finds-offshore-risk-management-and-regulatory-oversight-still-inadequate-in-gulf-of-mexico/

Douglas, M. (1992). *Risk and Blame: Essays in Cultural Theory*. Routledge.

Fruhen, L. S., Mearns, K. J., Flin, R., & Kirwan, B. (2014). *Safety Science*, *69*: 29–36.

Goodman, J. (2013). Rig's manager says BP tried to skip test, changed drilling plan. *Miami Herald*. www.idahostatesman.com/news/article40707309.html

Hammer, D. (2021). Jimmy Harrell, supervisor on doomed Deepwater Horizon. www.nola.com/news/environment/jimmy-harrell-supervisor-on-doomed-deepwater-horizon-drilling-rig-dies-at-65/article_4a7bf482-b368-11eb-bdf4-533ef352b155.html.

Hatfield, E., Cacioppo, J. T., & Rapson, R. L. (1993). Emotional contagion. *Current Directions in Psychological Science*, *2*(3): 96–99.

Hogeveen, J., Inzlicht, M., & Obhi, S. S. (2014). Power changes how the brain responds to others. *Journal of Experimental Psychology: General*, *143*(2): 755–762.

Hopkins, A. (2011). *Management Walk-arounds:Lessons from the Gulf of Mexico Oil Well Blowout*. National research center for OHS regulation. https://humanfactors101.com/wp-content/uploads/2022/07/management-walk-arounds-lessons-from-the-gulf-of-mexico-oil-well-blowout.pdf

Hopkins, A. (2012). *Disastrous Decisions: The Human and Organisational Causes of the Gulf of Mexico Blowout*. CCH Australia.

Hopkins, A. (2023). *The Human and Organisational Causes of the Gulf of Mexico Blowout*. Australian National University. Elgar Concise Encyclopedia of Oil and Gas Law.

IOGP. (2014). International Association of Oil and Gas Producers. https://humanfactors101.com/wp-content/uploads/20224/03/csb-macondo-investigation-report-volume-1.pdf

Kaluza, R., & Karinch, M. (2020). *Deepwater Deception: The Truth about the Tragic Blowout & Perversion of American Justice*. WN.

Long, R. (2012). *Risk Makes Sense: Human Judgment and Risk*. Human Dymensions.

McDaniel, R. R., Jr. (2007). Management strategies for complex adaptive systems: Sensemaking, learning, and improvisation. *Performance Improvement Quarterly*, *20*(2): 21–42.

McDaniel, R. R., Jordan, M. E., & Fleeman, B. F. (2003). Surprise, surprise, surprise! A complexity science view of the unexpected. *Health Care Management Review*, *28*(3): 266–278.

Pupulidy, I. (2015). *The Transformation of Accident Investigation: From Finding Cause to Sensemaking*. Published Dissertation, Tilburg University.

Pupulidy, I., & Vesel, C. (2023). *Human & Organization Potential*. Dynamic Inquiry LLC.

Reader, T. W., & O'Connor, P. (2014). The Deepwater Horizon explosion: Non-technical skills, safety culture, and system complexity. *Journal of Risk Research*, *17*(3): 405–424.

Roberts, R., Flin, R., & Cleland, J. (2015). Everything was fine: An analysis of the drill crew's situation awareness on Deepwater Horizon. *Journal of Loss Prevention in the Process Industries*, 38: 87–100.

The President's National Commission on the BP Deepwater Horizon Oil Spill and Offshore Drilling. (2011). *Deep Water: The Gulf Oil Disaster and the Future of Offshore Drilling*. Public Affairs. www.govinfo.gov/content/pkg/GPO-OILCOMMISSION/pdf/GPO-OILCOMMISSION.pdf

World values survey. (2023). WVS Cultural Map: 2023 Version Released. www.worldvaluessurvey.org/WVSNewsShow.jsp?ID=46744

11 The Counterinsurgency

Introduction

The following case study retells a true story of a safety manager nameDarren Brooks who created a genuine safety evolution.

The Counterinsurgency (COIN) Safety Strategy

> Our journey began with a core philosophy: Relationship-centered Safety ... It is not just about rules, but about human connection, trust, and shared understanding. We studied Kuhn (1996) to consider studies of how changes to paradigms occur because we believe the safety field is in the process of a paradigm change, and we want to be on the front line of that change. We also read Ralph Stacey's work on complexity and relationships (2003). I believe we're getting there now that we've cut accidents in half and we have the best relationships between employees, management and us that we've ever had.
>
> *Darren Brooks, creator of the Safety COIN strategy*
> *Opticore-Tech is a fictional name. The events are real*

COIN and Relationship-Centered Safety (RCS)

The COIN safety model draws inspiration from the strategies that Generals David Petraeus and James Mattis successfully employed in dynamic operational environments. The COIN doctrine emphasizes protecting the population rather than just killing the enemy, which is parallel to protecting the workforce vs. enforcing compliance. This people-centric safety approach, directly mirroring the core tenets of successful COIN operations, prioritizes winning the hearts and minds of the workforce. It seeks to foster a deep sense of trust, and mutual respect as the foundation for open communication.

An effective COIN doctrine underscores the critical importance of deeply understanding and collaboratively working with the local population to achieve sustainable security. As a safety model it seeks to win the support of employees, recognizes that the individuals on the front lines of work processes possess

DOI: 10.4324/9781003668770-11

invaluable, firsthand knowledge and critical insights into existing workplace hazards and the most practical and effective solutions for mitigating those risks.

The story of Opticore-Tech, a company comprising 800 employees, provides a powerful and tangible illustration of this safety COIN in practical action.

Quiet Surround Strategy

Opticore-Tech embarked on a carefully orchestrated quiet surround strategy over nine months, subtly and systematically shifting its traditional safety philosophy away from grand, top-down announcements toward an organic and relational transformation. Their strategic approach mirrored the COIN principle of facilitating new pathways for change, focusing on building momentum through genuine connection and collaborative problem-solving, recognizing the inherent resistance to sweeping organizational changes.

Per Brooks, resistance in this context is the workforce rejecting the organization's attempts to control the process. To prevent that the safety reps became safety facilitators and stop work became "I have a question." The ride alongs helped the facilitators learn the mental models of the workforce, which allowed them to understand at a deeper level the language the frontline used and ultimately understand their mental models and language. It created shared ownership and true collaboration in safety rather than disconnects.

A cornerstone of Opticore-Tech's strategy was the deliberate act of embedding safety facilitators directly within operational work groups. Significantly, the company made a conscious decision to rename safety technicians to safety facilitators. This seemingly simple change in terminology signaled a crucial shift in organizational mindset, moving away from a perception of safety personnel as enforcers with all the knowledge and towards a culture of collaboration, support, and shared problem-solving.

This subtle change in language demonstrated a deep understanding of the principle that words carry significant meaning and can either inadvertently create combative and adversarial dynamics between safety personnel and the workforce or, conversely, actively foster an environment of trust, open communication, and mutual respect.

Mirroring the COIN doctrine's profound emphasis on thoroughly understanding the operational environment from the perspective of the local population, Opticore-Tech trained their safety facilitators for nine weeks to actively engage with frontline workers through a practice they termed Ride Alongs. These ride alongs were not opportunities for the facilitators to evaluate or audit the operators' performance. Instead, their primary purpose was to allow the facilitators to immerse themselves in the daily work processes, to learn the specific language and terminology used by the workers, and to understand the inherent challenges and potential hazards directly through the operators' own lived experiences.

Achieving a thorough understanding of the operational environment from the perspective of the local population required daily structured activities. As Brooks noted, "We made it a point to go out with workers one at a time to learn their language so that we could talk about hazards and preventative measures in the way that makes sense to them." This direct, person-to-person engagement served to build strong foundations of trust and rapport between the safety facilitators and the workforce, enabling the facilitators to gain a deep and nuanced understanding of the specific challenges, constraints, and needs of each work group firsthand.

These actions align with the principles of social constructionism where mutual understanding is key for collaboration. Changing the name to facilitators and the ride alongs align perfectly with the fundamental principles of social constructionism, which posits that learning and action are inextricably intertwined and lead to shared understanding, which is crucial for effective collaboration and problem-solving.

Transforming Traditional SWA

A short time into their journey they had cut their accident rate by 50% and then Brooks decided to make SWA their keystone habit and tool for transformation. The catalyst was a single, alarming event. Two employees, unsure of which valve to turn during a critical operation, decided to flip a coin to choose. When Brooks discovered this, he realized that the existing SWA policy was a failure. The workers felt it was safer to gamble on a coin toss than to stop the job and ask for help.

Brooks realized that traditional SWA policies fail because they are intimidating. They often require paperwork, justification, and the social awkwardness of halting production. To fix this, Brooks removed the fear. He rebranded SWA into a simple, non-confrontational phrase: "I have a question."

This normalized the act of pausing. It allowed workers to address small things—like asking for help lifting a heavy bottle—without triggering a formal investigation. Brooks remarked, "Saying 'I've got a question'… is an act of team solidarity."

Per Brooks, one of the main reasons the SWA doesn't work is that management assumes that employees inherently understand job safety. Often, these policies are unclear about when and how to use them. They also require intimidating documentation that makes employees feel they must justify stopping work. Additionally, people focus on major, obvious dangers, missing the smaller, everyday safety issues. This last point was crucial. Brooks wanted SWA to feel like a normal part of operations, not just about halting work but about encouraging questions, which significantly shifted perceptions about SWA. These perceptions have also been documented by researchers (Greenshields, 2017; Weber et al., 2018).

Many point out that a lack of psychological safety prevents the use of SWA. Brooks never used that term. He removed fear by normalizing stop work as a simple question about work, making it as easy and natural as saying, "Hold on, I have a question." This encourages the recognition of even small things, like asking for help to lift a heavy water bottle as valid stop work actions. It also cuts through the perceived complexity and fear associated with formal stop-work procedures, making it more accessible and less intimidating.

Brooks remarked,

> No one on the team wants an environmental release. No one wants someone to get hurt. So saying "I've got a question," regardless of the size or the severity. Size doesn't matter because you are contributing to an effort to protect the goals and objectives of that team.

Brooks had transformed asking a question into an act of team solidarity toward safety.

Opticore-Tech demonstrated that SWA can be far more than just a policy. It's a social achievement deeply rooted in strong relationships, open communication and in shared understanding. When it's rooted in a supportive organizational culture that champions learning, it proactively addresses the complexities of balancing safety and production. Brooks commented, "You've developed a common understanding of safety standards that doesn't usually exist in a company."

SWA as a Keystone Habit

The SWA policy became the keystone habit and the vehicle for change. Keystone habits are not individual habits. They are key operational targets that can directly influence and even create group norms within an organization or team (Weber et al., 2018; Tharaldsen et al., 2008; Greenshields, 2017; Collins & Porras, 1994). A habit becomes a deeply ingrained belief and behavior that shape how employees think, act, and interact, and constantly challenged the status quo in pursuit of better results (Collins & Porras, 1994). By promoting and reinforcing a keystone habit, leaders can establish expectations for how members should behave and improve overall performance. An example of *habit* is described in the Paul O'Neil Alcoa case study in Chapter 13 (Duhigg, 2012).

Demonstrating Value and Overcoming Mindsets

According to Brooks, success has been notable, with departments reporting a newfound appreciation for embedded safety facilitators; "if I tried to remove them today, I'd face resistance." Unlike companies where safety personnel and their value are unseen, at Opticore-Tech, managers and supervisors have come to rely on the daily activity reports that keep them informed of potential risks. Whereas, in the past, operations saw safety facilitators as mere helpers, they were now seen as essential to operations. "We had to break down old mindsets. Our managers now see how safety has improved their entire operation, and understand how asking a question before acting is a self-applied safety interaction that prevents incidents."

At the heart of Opticore-Tech's transformation was the Daily Activity Report (DAR). Rather than a standard compliance checklist, the DAR served as a strategic guide for the facilitators, ensuring that every shift was anchored in specific, interpersonal relationship activities.The report prioritized "High-Value Interactions" over administrative accounting. Key components included. It also served as a record of measurable activities that served as leading indicators.

- **Work Permit Observations:** Facilitators observed the issuing process from start to finish. This was not a "check-the-box" exercise; it was an opportunity for real-time coaching and mentoring, with every interaction documented to track the growth of the relationship.
- **Ride Alongs:** Facilitators made rounds with operators to learn the plant's nuances. These sessions provided a natural environment for task-specific hazard assessments based on the workers' lived experiences. At the end of the ride, a daily activity sheet was shared with the manager to go over concerns, needs, and things that were going well. These reports provided management with operational intelligence (e.g., "The crew is struggling with X tool") rather than just safety statistics. This made the safety team essential to operations.
- **Job Safety Observations (JSO):** Facilitators checked every maintenance job daily, ensuring safety wasn't a distant policy but a present, collaborative effort.
- **Employee Consults:** This functioned as an informal "Open Door" policy. If an employee had a question regarding health or safety, the facilitator documented the concern and either addressed it immediately or scheduled a formal follow-up, ensuring no voice was ignored.

To move the culture away from a retributive mindset, the team purged the words "audit" and "inspection" from their vocabulary. For high-risk activities like Lockout/Tagout (LOTO) and Confined Space Entry, they replaced bureaucratic checklists with a simple, binary communication: "Go" or "Stop." This was a direct application of Thomas Kuhn's (1996) paradigm shift—moving from a world of individual blame to a world of collective, relationship-centered achievement.

Outcomes and Lessons Learned

Opticore-Tech's safety COIN provides compelling real-world evidence that adopting an RCS philosophy can lead to significant reductions in accidents and collaboration around safety. The results demonstrate the power of building trust and open communication through relationships; the impact of language on safety and relationships; the value of understanding lived experiences; the effectiveness of leading indicators related to relationships and well-being; and the success of a gradual, relationship-focused approach to change.

Summary of Opticore-Tech's RCS Improvement Framework

Opticore-Tech's approach (Figure 11.1) involved the deliberate creation and implementation of repeatable activities such as the ride-alongs specifically designed to foster strong interpersonal relationships and sustain consistent engagement between safety personnel, supervision, and the workforce. So, people began experiencing trust and open communication in their normal work exchanges.

If the RCS program were to be represented in a summary model, it might look like Figure 11.1.

> This cycle operates on the premise that once a relationship is established, people tend to want to maintain it. Combined, these activities form the structure of RCS:
>
> 1 Establish a relationship
> 2 Relationship leads to trust
> 3 Trust leads to commitment
> 4 Select a keystone habit that opens pathways for active communication.
> 5 Full circle to maintain relationships

Conclusion: A Paradigm Shift toward Human Connection

The Opticore-Tech safety transformation began with a framework of RCS and other key strategies such as Nemawashi and "I have a question." This case study offers a powerful testament to the power of relationships, demonstrating that a human-centric approach, akin to the military COIN strategy, can overcome the over-reliance on traditional, compliance-driven safety models. By prioritizing trust, open communication, and genuine connection, Opticore-Tech not only achieved a remarkable 50% reduction in accidents but also cultivated a deeply engaged and committed workforce.

The success of Opticore-Tech demonstrates that concepts like social construction and sensemaking are not merely academic but practical forces, shaping how individuals collectively understand and respond to safety within their shared work environment. The goal is fostering an environment where every employee feels valued and understands how their contributions increase collective well-being. The following chapter will delve into the specific tools and skills that underpin the Opticore-Tech approach, demonstrating how these competencies can be developed and applied to achieve a similar paradigm shift.

Executive Summary and Takeaways

The key executive takeaways, focused on the deliberate social construction of a safer reality, are:

Strategically use language to shift underlying power dynamics. Opticore-Tech achieved this by renaming safety personnel from "safety technicians" (perceived as enforcers) to "safety facilitators". This subtle yet powerful linguistic shift redefined the relationship from one of command-and-control to one of collaboration and mutual respect.

Risk is a subjective, socially constructed concept, not solely objective data. To gain accurate, ground-level awareness, leaders must mandate processes that value

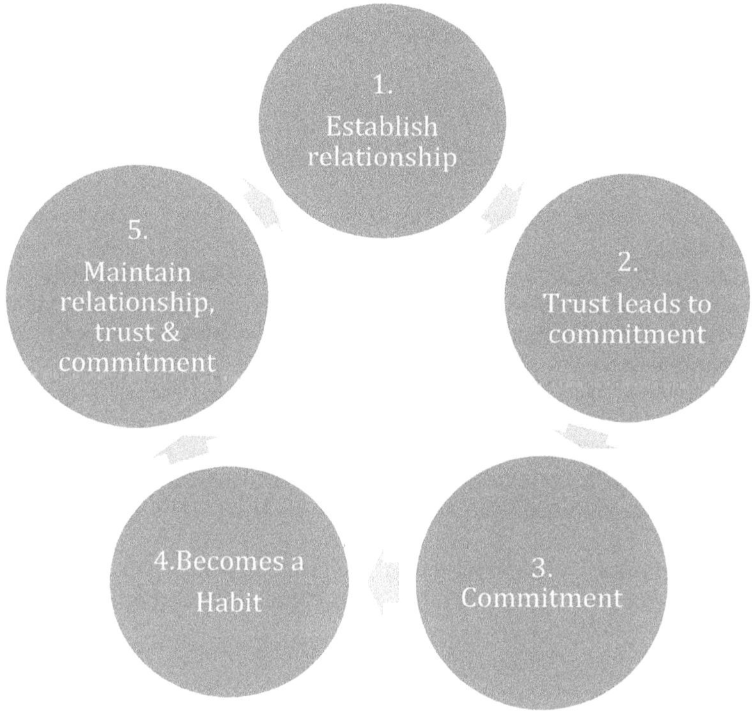

Figure 11.1 Opticore-Tech's RCS improvement framework.

the worker's reality. The "Ride Alongs" were crucial repeatable activities designed specifically to immerse safety facilitators in daily work, learn the workers' language, and understand hazards directly through their lived experiences. This direct, person-to-person engagement aligns with social constructionism, where mutual understanding leads to effective collaboration and shared risk awareness. It anchors risk management in shared meaning and lived experience.

Operationalize Psychological Safety by Eliminating the Need to "Save Face": The fear of embarrassment or loss of status—the need to "save face"—is the primary social barrier that silences employees. Opticore-Tech intentionally overcame this psychological obstacle by reframing SWA, transforming it into the non-confrontational phrase, "I have a question". This action successfully socially constructed speaking up as an act of team solidarity toward protecting collective goals, thereby aligning the employee's self-interest (Belief #5) with the organization's interest in stopping unsafe work.

Validate RCS as the Blueprint for Collective Achievement: The Opticore-Tech success provides tangible, real-world evidence that adopting an RCS philosophy—built on core generative beliefs and implemented through intentional social actions—results in significant reductions in accidents and a highly engaged workforce. The model demonstrates that safety is a collective achievement and an emergent property of a system founded on trust, commitment, and open communication.

References

Collins, J. C., & Porras, J. I. (1994). *Built to Last: Successful Habits of Visionary Companies*. HarperBusiness.

Duhigg, C. (2012). *The Power of Habit: Why We Do What We Do in Life and Business*. The New York Times.

Greenshields, H. (2017). *Why Formal Stop Work Schemes Do Not Work: A Qualitative Study in Safety-Critical Industries in Canada.* Lund University of Sweden. https://safetyinsights.org/2024/08/15/why-formal-stop-work-schemes-do-not-work-a-qualitative-study-in-safety-critical-industries-in-canada/.

Kuhn, T. S. (1996). *The Structure of Scientific Revolutions*. 3rd Ed. University of Chicago Press.

Stacey, R. (2003). *Complex Responsive Processes in Organizations*. Routledge.

Tharaldsen, J. E., Olsen, E., & Rundmo, T. (2008). A longitudinal study of the safety climate on the Norwegian continental shelf. *Safety Science*, 46(3): 427–439.

Weber, D. E., MacGregor, S. C., Provan, D. J., & Rae, A. (2018). "We can stop work, but then nothing gets done." Factors that support and hinder a workforce to discontinue work for safety. *Science Direct*, 108: 149–160. https://forgeworks.com/wp-content/uploads/2022/09/Authority-to-Stop-Work.pdf

12 Implementing Relationship-Centered Safety

This chapter translates the theoretical framework of relationship-centered safety (Carrillo, 2019) into a practical blueprint. Using the Opticore-Tech case study, we detail the specific, repeatable actions—the "micro-changes"—that countered the effects of individualism and built genuine collective responsibility.

Success stories, like Opticore-Tech and Rio Tinto, provide the clear blueprint for transforming this philosophy into action, demonstrating that intentional social design leads to measurable organizational improvements.

Counterinsurgency (COIN) Implementation Timeline

The Opticore-Tech implementation was structured as a "quiet surround strategy," moving systematically from internal preparation to gradual implementation across the organization, mirroring the principles of a safety COIN. The purpose of this nine-month quiet phase was to allow micro changes time to integrate knowledge with experience and ownership, thereby building the relationships necessary to sustain new behaviors.

The *safety COIN* model utilizes: (1) a stop work authority (SWA) policy, (2) embedding safety facilitators directly into operational work groups, and the (3) safety facilitator's DAR which serves as a way to track results. In addition, it intuitively employed the principles of social construction and sensemaking to change the culture (Greenshields, 2017).

Operationalizing Key RCS Components

Renaming and Restructuring (Action 1: Change the Social Dynamic)

Opticore-Tech renamed safety staff from "safety technicians" (perceived as enforcers) to "safety facilitators" (collaborators). This deliberate linguistic shift was subtle yet powerful, redefining the relationship from one of command-and-control to one of collaboration and mutual respect. This action aligns with Belief #2: Inclusion precedes accountability.

DOI: 10.4324/9781003668770-12

Table 12.1 Implementation timeline

Implementation phase	*Timeline*	*Key activities and milestones*
Quiet Preparation (Phase 1)	Nine months	The team quietly prepared for the philosophical shift, utilizing a "quiet surround" strategy to let change integrate organically and mitigate the anxiety often associated with new programs.
Gaining Approval (Phase 2)	Strategic point	The site safety team presented the plan to management, strategically leveraging Kuhn's (1996) paradigm shift model and the concepts of relationship-centered safety (RCS), Carrillo (2019).
Restructuring and Language Shift (Phase 3)	Immediate action	Renamed "safety technicians" to "safety facilitators" to establish a relationship of respect. Abstract or combative terms like "Audit" and "Inspection" were replaced with neutral, collaborative language like "Ride-Along".
Facilitator Training (Phase 4)	Nine weeks	Safety facilitators received intensive training to actively engage with frontline workers, learning their language, and job challenges (not auditing performance).
Repeatable Relationship Activities and SWA Rollout (Phase 5)	Ongoing Daily	Began daily, one-on-one "Ride-Alongs". Formalized the non-confrontational phrase "Hold on, I've got a question" as the official Stop Work Authority (SWA) method.
Outcome	Long-term	The relationship-centered philosophy successfully countered the effects of individualism, leading to a demonstrable 50% reduction in accidents.

Department Management Meetings and Nemawashi

The inclusion of safety facilitators attending department management meetings was crucial; while software can manage action items, it cannot replace person-to-person conversations for building trust and communication. These were carefully managed,

> When doing things, I follow the Japanese process of Nemawashi—small discussions with individuals or small groups before a meeting occurs. Decisions are basically agreed upon prior to a meeting, then the meeting simply serves as a process to ratify a proposal. Very effective and probably a key practice in our success.

The Result: The public meeting becomes a process for ratification rather than a battleground. This saves "face" for managers, prevents public conflict, and ensures that when a safety decision is reached, it already has the quiet support of the entire leadership team.

Repeatable Relationship Activities

Opticore-Tech established a series of monthly, one-on-one activities with employees designed to foster and maintain relationships between them and safety facilitators. As detailed in the Opticore-Tech case in Chapter 11, the team utilized five facilitators to execute a specific rhythm of interpersonal activities, including permit coaching and safety patrols.

> We put 5 safety facilitators in our operating departments, each completing a DAR that guides their day and creates very specific interpersonal relationship activities. For example, each day they do work permit issuing observations with coaching and mentoring documented in detail. They also perform ride-alongs for task-specific hazard assessment.

These carefully structured one-on-one activities produced information that helped managers improve operations. The information was also used to promote ongoing, significant dialogue between safety facilitators and the workforce. The safety manager holds a weekly safety facilitator meeting to review activities, target support resources, and debrief on departmental safety status. These activities began shaping the work environment. We saw more trust and open communication.

The Daily Activity Report (DAR)

Trust does not happen by accident; it must be engineered. Opticore-Tech established a mandatory **Daily Activity Report** for facilitators. The DAR is not a checklist of hazards, but a record of high-value social interactions. For implementation, the DAR serves as the primary mechanism for turning 'soft' social data into 'hard' operational intelligence, identifying systemic signals that traditional audits miss."

The report prioritized "High-Value Interactions" over administrative counting. Key components included:

Permit Coaching:

- **The Ride-Along:** A specific daily requirement to shadow a task, not to judge it but to understand it.
- **The Output:** These reports provided management with operational intelligence (e.g., "The crew is struggling with X tool") rather than just safety statistics. This made the safety team essential to operations.

Stop Work Authority (SWA): "I've got a question"

A year into implementation, the team realized that the traditional SWA was failing because of fear. They reframed SWA as a **Keystone Habit**.

The New Protocol:

- **The Phrase:**"Hold on, I have a question."
- **The Logic:** Stopping work is intimidating; asking a question is natural. This radical simplification removed the bureaucracy and fear associated with halting production.
- **The Scope:** It recognizes that asking for help (e.g., lifting a heavy box) is a valid SWA event. It is an act of team solidarity, not a whistleblowing event.
- **Implementation:** Management was trained to welcome these questions without penalty. Every "question" was recorded in the Daily Activity Report as a positive safety contribution, reinforcing the behavior.

Implementing "I've got a question."

A Collaborative Tool, not a Punishment

The focus shifts from individual compliance to collective responsibility. Everyone on the team is responsible for each other's safety and the overall goals of the project.

Encourage Questions, Don't Intimidate: SWA should be as simple as asking a question– a radical simplification that removes the fear and bureaucracy often associated with stopping work. This makes it a normal part of daily operations.

Recognize All Safety Contributions: This program recognizes that even small actions, like asking for help with a heavy object, are valid uses of SWA. It's not just for major, obvious dangers, but for every small issue that could potentially lead to an incident.

Clear and Accessible Policies

Eliminate ambiguous rules and intimidating documentation."

Build Relationships and Trust: The program is built on the foundation of strong relationships and open communication. When trust exists, people are more likely to speak up without fear of negative consequences. This fosters a sense of psychological safety.

Align Safety with Team Goals: Frame SWA as an act of team solidarity. Stopping work to address a safety concern protects the team's objectives and ensures everyone can go home safely.

Simplify the Process: Formalize the phrase, "Hold on, I've got a question," as the official way to use SWA. Eliminate any intimidating paperwork or formal procedures for small-scale stops.

Train for Everyday Situations: Training should focus on recognizing both large and small potential risks. Use real-world examples from the team's daily activities to illustrate how SWA can be applied. It can be as simple as asking for help to lift a drinking water bottle.

Promote Open Communication: Management must visibly support SWA by welcoming questions and treating every stop-work action as a valuable contribution. Never penalize an employee for using SWA.

Make SWA Contributions Visible: Use tools like Daily Activity Reports to highlight and celebrate safety contributions, no matter how small. This reinforces the value of SWA and the person who used it (Weber, et al., 2018).

Foster a Common Understanding: Regularly discuss safety standards and expectations as a team. This ensures that everyone shares a common understanding of what is and isn't safe, fostering a collective conscience.

RCS Eight Beliefs Demonstrated at Opticore-Tech

How Opticore-Tech Operationalized the eight RCS Beliefs: The success at Opticore-Tech wasn't magic; it was the mechanical application of the RCS Beliefs:

- Belief #1 (Communication requires relationship): The Ride-Alongs ensured that communication happened in the context of a relationship, not an inspection.
- Belief #2 (Inclusion precedes accountability): Renaming staff to Facilitators signaled that inclusion was the priority. You cannot hold someone accountable if you haven't first helped them.
- Belief #4 (People want to contribute): The Daily Activity Reports highlighted small wins, proving that workers were willing to contribute when given a path to do so.
- Belief #5 (Self-Interest): Reframing SWA to "I have a question" aligned safety with the worker's self-interest (avoiding confusion) rather than the company's interest (avoiding lawsuits).
- Belief #8 (Relationships influence decisions): The use of Nemawashi acknowledged that decisions are emotional. By managing the relationship *before* the meeting, they managed the decision *during* the meeting.

Conclusion

The Opticore-Tech case serves as a practical demonstration of how OHS advisors can successfully implement change by utilizing relationship-centered concepts, such as riding along with employees to build trust. They implemented a relationship-centered philosophy that led to significant reductions in accidents through building trust and open communication. Without this foundation of trust, communication remains merely a transaction, and crucial safety tools, such as SWA, fail because the necessary social and emotional conditions such as lack of fear do not exist.

The final mandate for leadership is clear: a transition from a top-down, command-and-control approach to one that trusts on-site employees to make informed decisions. By valuing the intrinsic human quality of generativity and nurturing the social fabric, organizations build a healthier, more adaptable, and truly resilient enterprise.

References

Brooks, D. (Source of the safety counterinsurgency strategy, quoted in Chapter 11).

Carrillo, R.A. (2019). *The Relationship Factor in Safety Leadership: Achieving Success through Employee Engagement*. Routledge.

Greenshields, H. (2017). *Why Formal Stop Work Schemes Do Not work: A Qualitative Study in Safety-Critical Industries in Canada*. Lund University of Sweden. https://safetyinsights.org/2024/08/15/why-formal-stop-work-schemes-do-not-work-a-qualitative-study-in-safety-critical-industries-in-canada/.

Kuhn, T. S. (1996). *The Structure of Scientific Revolutions*. University of Chicago Press.

Weber, D. E., MacGregor, S.C., Provan, D. J., & Rae, A. (2018)."We can stop work, but then nothing gets done." Factors that support and hinder a workforce to discontinue work for safety. *Science Direct*, 08: 149–160. https://forgeworks.com/wp-content/uploads/2022/09/Authority-to-Stop-Work.pdf.

13 OHS Risk Navigator

Introduction

> "If we don't navigate uncertainty, well… people get hurt. To do it well you need certainty in your relationships with each other and with uncertainty."
>
> Laurin Mooney

What is an occupational health and safety (OHS) advisor's role? Some progress has been made; however, it is still often perceived as poorly defined and sometimes adversarial (Provan et al., 2018). The majority of OHS professionals feel undervalued, unheard, excluded, and nearly 53% contemplate leaving their positions (Carrillo, 2024). Overwhelmingly, in my interviews with advisors they reveal that they are not able to speak freely. Managers don't listen and "you can only push so much." The degree of emotional labor required to be in these positions is unacceptable.

Yet, the vast majority show a passion for their work and a willingness to struggle with the existential conflicts that surround them daily. When speaking about their work, practitioners used words such as caring and helping. These are not terms found in any definition of OHS by associations, universities, or government agencies. In fact, they are words associated with a helping profession. In the words of an OHS advisor,

> I often wonder what really drives us in this profession. When people say they care what does that mean? I know for me it's about understanding people, learning about the process through them and really connecting with them on a human level.
>
> (Personal conversation with OHS advisor, 2025)

Champion a Generative, Human-Centered and Caring Approach

This is the core philosophy of RCS. It is the "internal compass" that guides the Risk Navigator. This ethic provides the intrinsic motivation for all other actions and aligns with the call for a human-centered approach. The value of this approach

DOI: 10.4324/9781003668770-13

was demonstrated by the GM for the Rio Tinto Mine when he instructed all his managers to look employees in the eye, use their name and say hello.

These kinds of generative actions are within the power of OHS advisors. And it fits the role because ultimately, many safety professionals act as the conscience of the immune system, constantly advocating for its health and countering the "autoimmune disease" where production pressures attack the organization's own well-being.

While advisors can play a critical role in communicating company care for employees, organizations must ensure this doesn't become an undue burden on them. Often, they already face limited resources and competing priorities. Managers can support advisors by communicating that a people-centered approach is a shared responsibility, not solely the burden of safety advisors.

These skills fit OHS practitioners because they identify as being in the "helping professions," driven by a moral concern for worker well-being. This aligns with the concept of Generativity—the motivation to guide, support, and care for others and future generations. This generative drive provides the powerful "why" for their work, moving it beyond compliance to a genuine commitment to people.

Despite the consistent use of the term OHS advisor throughout this book, there is very little agreement on a title for those who practice health and safety. Nor is there agreement on their job description (INSHPO, 2017). To explore this issue further, a group of OHS advisors attended a mind mapping session to identify their responsibilities, the roles they play, and the activities they perform (see Figure 13.1). Through this activity they discovered that they spend 67% of their time on soft infrastructure building (see Figure 13.2).

Sixty-seven percent of OHS advisors' responsibilities require soft skills (people skills). Thirty-three percent of responsibilities require technical or "hard skills."

Participants felt the mind map was enlightening. One commented, "This is the first time I've understood the job I'm doing and why I feel overwhelmed." Another said, "I feel helpless because I'm unable to improve any of these situations." And a third: "No wonder I've never been able to explain what I do when people ask me about my job." The burnout and exhaustion they describe is the emotional cost of trying to fulfill this 67% relational role without the requisite skills.

Advisors are indeed navigating a sea of unpredictable conflicts, emotions, fears, envy, and power plays. The words used by advisors in interviews—burnout, exclusion, and contemplating resignation—are the physical and emotional costs of trying to operate in those chaotic waters without the mandate or tools to navigate them collectively. Risk navigator emerged as an appropriate job description for people who are on the frontline mitigating physical and emotional risks.

High job demands accompanied by a lack of resources can lead to angst when advisors cannot effectively address the people issues they face. Recognizing the systemic nature of these challenges can offer emotional protection. As one advisor said, "This mind map reflects my reality, but I never knew how to articulate it. At least I know now that I'm not crazy." They felt validated by the exercise and their boss followed up with many conversations about the support they needed.

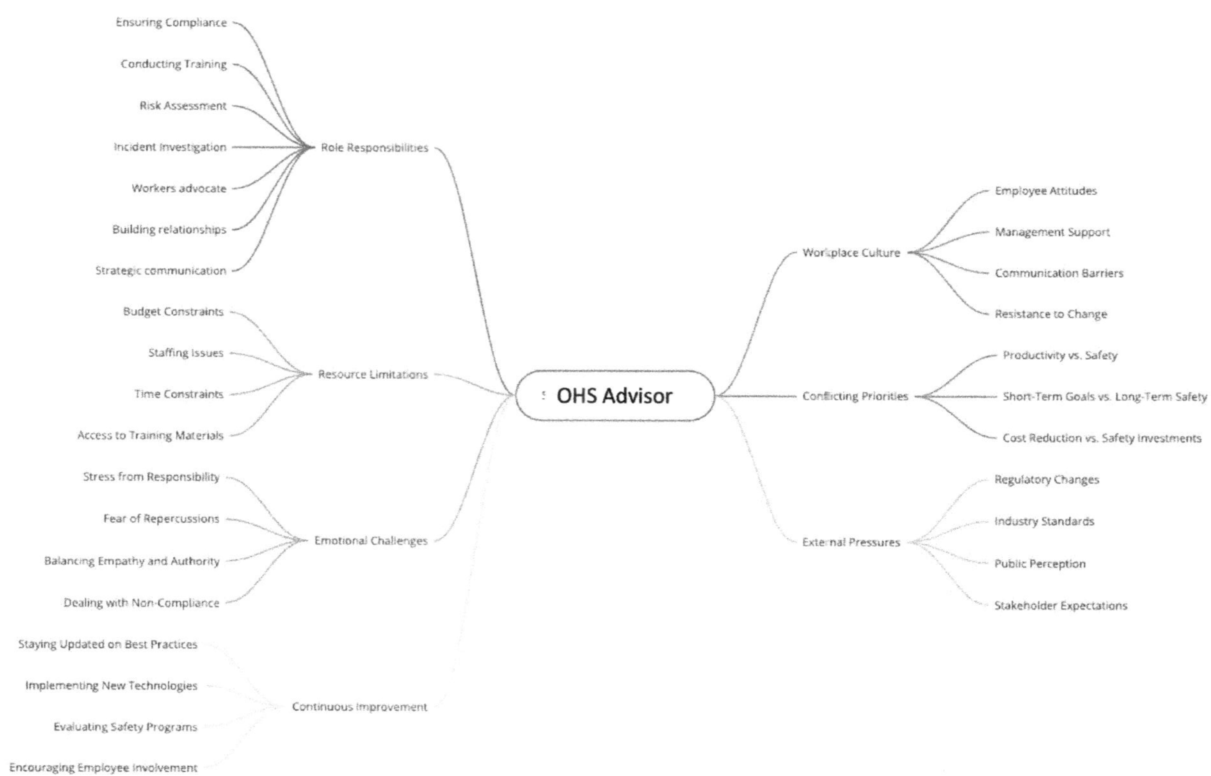

Figure 13.1 OHS advisor’s lived experience.

Source: Carrillo (2025).

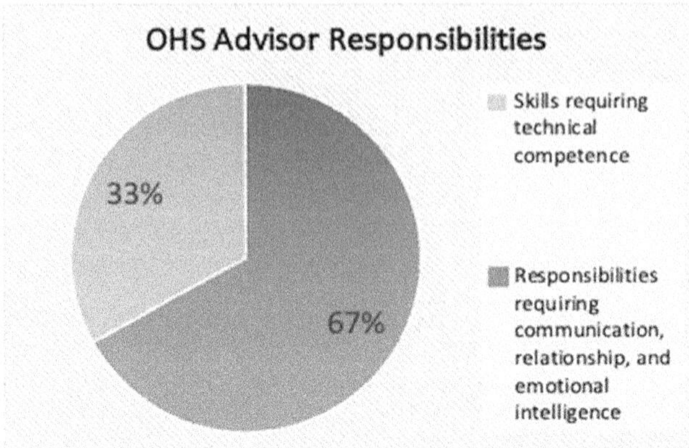

Figure 13.2 Sixty-seven percent of OHS advisors' responsibilities are in the soft infrastructure category.

The isolation of their role is a constant struggle. In another study (Provan et al., 2018), an advisor exposed the waste of time and talent:

> I would always in passing hear about a meeting because no one else was at their desks. So, the structural guys aren't there, mechanical guys aren't there, piping isn't there, project managers aren't there – safety are all sitting in their cubical [...] and we're thinking we're making this place safer, but little did we know, upstairs the whole entire design team, except safety, are sitting there making decisions that are contrary to the documents we're writing
>
> (25)

Some advisors have created strong positions for themselves within a company. However, the majority feel they are on their own without visible support. The discussion also made it clear that they are hampered by outdated safety paradigms where management lacks the understanding or the motivation to invest more in the people side of safety. A fundamental shift toward a more human-centered, relationship-based, and psychosocially informed approach can help them meet the challenges in the workplace. Changing their identity from advisor to risk navigator provides a more comprehensive understanding of their responsibilities and the skills they need to be successful.

While the "living system" metaphor primarily guides executives in recognizing and investing in an organization's "social architecture," the "OHS advisor as risk navigator" metaphor focuses on the individual practitioner's role and challenges within a complex organizational environment. It draws parallels to ancient mariners using rudimentary tools and intuition to navigate unpredictable waters, just as the advisor must negotiate the social seas.

A New Professional Blueprint: The Metaphor of Navigating Risk

The state of navigation in the 15th–18th centuries shares commonalities with risk management today. That period, the Age of Exploration, was a perilous affair characterized by significant uncertainty and rudimentary tools. Yet people went ahead, exploring the unknown using intuitive tools that we may have forgotten and might need to resurface.

By reframing risk management as navigating risk, all are alerted to drop any pretense of having everything under control. Weather and conditions can change at any moment; the "dead reckoning" records left by other sea captains could have been mistaken. Consider the terror of being lost for weeks, the sun and stars their only guides, the gnawing anxiety of a silent, windless sea as food and hope ebbed away. These were not operational disruptions; they were existential crises. The "dead reckoning" they employed, a constant, imperfect calculation based on feel and guesswork, speaks to a profound reliance on intuition, on an embodied understanding of the ship and the sea that modern technology often screens us from acknowledging.

Reliance on Embodied Knowledge

The seasoned mariner developed an almost preternatural sensitivity to the subtle shifts in the wind's breath, the barely visible changes in the ocean's currents. The same talent exists in the OHS advisor—a similar, vital awareness. This receptivity to the intangible (embodied knowledge) was the ancient sailor's lifeline to the physical world. For the modern advisor, their lifeline is awareness of the subtle signs of human emotions, the shifts in trust or morale, the unspoken anxieties rippling through the workforce. These are the "winds and currents" of the modern workplace, the precursors to strikes, safety violations, and systemic failures. For them, the social landscape of the organization is equally powerful to the ocean.

Perhaps the ancient mariner forced to rely on an intimate, almost symbiotic relationship with their environment, possessed a form of risk intelligence we are in danger of losing, especially when we consider our increased reliance on technology. Their survival wasn't just about reacting to the visible; it was about sensing the invisible. This skill is honed by constant awareness and a deep respect for forces beyond their complete control. In our increasingly data-driven world, are we cultivating this same intuitive awareness of the human element? Or are we too busy looking at a screen, blinded to the subtle, social signals that provide the wisdom for navigating the inherent uncertainties of our complex workplaces?

I am using the metaphor of the OHS advisor as a navigator. Like the ancient mariner who went into the open sea without modern technology, the OHS advisor navigates a sea of unpredictable conflicts, emotions, fears, envy, and power plays. When dealing with people, there is no technology that can tell you when you're going off course or about to hit a metaphorical iceberg. People's emotions and past experiences are invisible to the eye, so one can never be sure of how their words are being interpreted. Thus, there's always the possibility of cutting off the communication pathways with the crew and losing the vital early-warning sensors.

Bringing the Unseen into View

The "safety iceberg," a symbol of unseen dangers, also fits into the metaphor of risk navigator. Similarly, for the 18th-century navigator, the intangible signs of approaching danger often lay beneath the surface. While advisors are trained in the physical aspects of accident prevention, safety and well-being are also significantly impacted by intangible emotions, trust, feelings, and relationships. It is essential to recognize, understand, and address these intangible risks to move beyond a superficial approach to safety. The first step in gaining a deeper understanding of these human characteristics is by pursuing greater self-awareness and expanding intuition.

One advisor wrote,

> One of the challenges we face in today's complex work environment is the abstract nature of information we share to our frontline employees. Traditional safety models often make us believe that work is linear and straightforward. This creates the illusion that as long as employees follow the rulebook, everyone will go home safe. While these rules are well-intentioned and vital, they often fail to connect with the reality of how work is actually done, which is often complex and highly adaptive. I think using employee language can bridge the gap between abstract safety information and the real-life experiences of frontline employees in their complex, messy, unpredictable, and highly adaptive work environment.
>
> (Tetteh, 2025)

Essential Skills for Risk Navigation

A complete list of suggested roles, responsibilities, and skills is available in Appendix A, Risk Navigator Job Description.

Just as ancient mariners relied on intuition and awareness of their environment, modern OHS advisors need to cultivate an awareness of the "social landscape" in their organizations, including emotions. Effective risk navigation requires soft infrastructure skills, including understanding and influencing human behavior, navigating organizational pressures, and fostering communication and trust.

This collective effort to identify a purpose for OHS is in alignment with RCS and organization as a living system.

The modern OHS executive role within the immunization network is that of social architect, risk navigator, and relationship builder. They are less of a compliance officer and more of a coach, facilitator, and strategic advisor whose primary

Exhibit 13.1 High-quality and low-quality conversation characteristics

Low Quality:

- Transactional: Data transfer only.
- Ambiguous: "Be careful out there."
- Silencing: "Don't bring me problems, bring me solutions.

High-Quality:

- Generative: "What are you seeing that I'm not seeing?"
- Committed: "I request that we stop until X is fixed."/"I promise to fix X by noon."
- Connecting: "I want to understand your experience with this tool."

responsibilities are to nurture the social health of the organization and strengthen the soft infrastructure. In this role they are masters of sensemaking and conversation.

As described in Chapters 7 and 9, an organization is a network of relationships, conversations, and shared practices that truly determines the organization's health and resilience. If an organization is a conversation, then the Risk Navigator's job is to monitor the quality of that conversation. Exhibit 13.1 defines the difference between low- and high-quality conversations.

OHS advisors point out that they have limited influence over important factors like production pressure or workforce capacity. One area where they can have a significant impact is through communication and relationship building. They can educate leaders and managers about the impact of social and psychological factors on safety outcomes, encourage the adoption of more inclusive communication practices, and promote a culture where everyone feels valued and heard.

The Opticore-Tech case demonstrates how OHS advisors can do all these things. They proposed and implemented a new structure utilizing relationship-centered concepts such as riding along with employees to learn about their challenges and their jobs. The full story is in Chapter 11, The counterinsurgency.

Collective Sensemaking and Learning

Collective sensemaking and learning from accidents includes the proactive work of helping teams build shared awareness before an incident and to understand the socio-technical influences on the workers after an accident. This skill directly incorporates the lessons from Deepwater Horizon's failure in sensing and learning. It positions the advisor as a facilitator of the social construction and sensemaking processes that the book champions.

Saving Face and Navigating Influence without Power

These are the specific, how-to skills used in daily interactions to leverage influence and relationship building. Saving Face is the key communication tactic for building psychological safety. *Navigating Influence* is the core political skill for operating within the organizational structure.

Saving face stands out as critically important. The quickest way to shut down communication is to embarrass someone or make them feel like they've failed. This principle applies universally, whether communicating with a direct report or a superior (Goffman, 1959). While many focus on this when communicating upwards, it's often overlooked when interacting with peers or those perceived as lower in the hierarchy. This oversight is a significant mistake, as, in challenging environments, every team member's alliance is crucial. Opticore-Tech demonstrated how to do this by reframing their SWA program as a request for information.

Experts emphasize creating an environment of psychological safety to enable all team members to help identify early warning signs of trouble. That doesn't mean that the goal is to avoid confrontation when needed. A proficient leader understands that a safe environment isn't one devoid of conflict, but rather one where conflict is managed without causing individuals to lose dignity.

Psychological safety is a group dynamic whereas saving face aids learning and open communication by preserving dignity and social standing, acting as an internal compass against humiliation. Essentially, psychological safety is the environment, while saving face is a skill to maintain connection, thus building a safer space. As important as it is to allow people to save face it can also be used as a reason to hide information. Someone might not want to admit they don't know how to perform a task for fear of losing status.

As personal anecdotes from safety advisors reveal, genuine communication—characterized by managers who care and are comfortable engaging with their teams—is highly valued by employees. This highlights the critical need for leaders to cultivate empathetic communication and a language of care specifically recognizing taxing emotional labor. The ability to go into the field, ask questions, and engage authentically provides access to accurate information about frontline realities that formal channels often miss. Conversation and relationship-building can help overcome the face-saving instincts to hide mistakes. This is proven by Opticore-Tech's safety facilitators who actively learned the workers' language to understand their lived experiences, thereby building the trust necessary for true communication.

A significant, often unspoken, aspect of this profession is how to operate effectively within a hierarchical environment when lacking formal authority or significant influence. There's a growing recognition of the complexities and potential downsides of constantly challenging the status quo within organizations. While "speaking up" is often encouraged, the reality can be taxing, demanding considerable emotional labor with limited immediate reward and potential personal risk. This constant pushback can contribute to burnout.

Effectively navigating this environment requires not only empathetic communication from leaders but also a strategic approach from advisors to overcome the inherent challenges of operating without formal authority, balancing the ideal of "speaking truth to power" with the realities of organizational hierarchy and personal well-being.

Navigating Internally to Manage Personal Stress and Improve Communication

While *saving face* is the essential social skill used to preserve the dignity of others and keep communication open, it must be balanced with *differentation* . If Saving Face is the compass that helps the Risk Navigator communicate with others, Differentiation is the internal shield that prevents the navigator from sinking under the weight of the organization's anxiety.

A valuable psychological dimension—*differentiation*—can increase the Risk Navigator's effectiveness and protect them from burnout. To survive the emotional labor of this role, the Risk Navigator must move from an undifferentiated state—thinking, "If I can just make them understand, I will feel better"—to a differentiated state. This means becoming an observer and holding yourself apart from responsibilities that belong to others. For example, if you perceive that your advice is ignored, saying: "I will state my professional truth clearly once. Whether they understand or act on it is their responsibility, not mine." This shift moves the advisor from pleading to boundary-setting. When facing resistance, advisors often fall into the trap of over-explaining to manage their own anxiety. To stop this cycle, avoid the trap of **JADE** : once professional facts are delivered, do not **Justify, Argue, Defend, or Explain**. Over-explaining is often a defense mechanism to manage one's own anxiety however in the face of resistance, it signals insecurity and provides more openings to attacks.

- **Example**: A supervisor snaps, "You have no idea how things actually work on the floor."
- **Undifferentiated Response (JADE):** "Actually, I spent three years in the field before I got my certification, and I've been on this floor every day this week. I'm trying to help you!"***Why it fails**:* It defends and justifies, which signals that the supervisor's comment successfully "hooked" the advisor's identity. You have dipped into negative emotional labor territory.
- **The Differentiated Response (Saving Face + Differentiation):** "You're right that I don't see everything you see. And I don't know everything you know. I know the technical safety requirements. From your experience where do you see these safety requirements clashing with the way the work actually gets done?"
- ***Why it works:*** It saves the supervisor's face by acknowledging their expertise, but remains differentiated by not taking the "desk jockey" insult personally. It pivots from an attack into a diagnostic question.

By staying in this differentiated space, the advisor becomes the "Calm Center" of the storm. They aren't trying to control the waves (the people or the events); their awareness and keen sense of social perception guides them to ask questions and respond to concerns.

Ultimately, the Risk Navigator must redefine success. As Carl Jung suggested, the goal is not perfection but wholeness. For the OHS advisor, wholeness means accepting that while you cannot control a manager's mind, you can influence the workers' environment. By withdrawing the energy spent on "shaping words" to please the boss, you reclaim the voice needed to speak for the people.

Case Study on Saving Face and Speaking Truth

This case study comes from my experience and demonstrates how some of these ideas apply in practice. I remember the first time I was scheduled to speak to a CEO. It was a nuclear business. The workforce had gotten an email informing them that they were being sent home for two weeks without pay after an alarm failed to function. The workers, most of whom lived paycheck to paycheck, were furious that they were being left to pay the penalty for something that was not their fault. Furthermore, all the supervisors on up were still working full-time and getting paid. The corporation was surprised when the employees began to organize protests and I was asked to come in and help to reopen communication.

I had interviews with employee groups, supervisors, and managers, then I was scheduled to meet with the CEO. I was told I had 15 minutes to explain why the workforce was in rebellion, the potential circumstances, and to recommend a response. I was also told not to bring up the topics of unfairness, double standards and lack of shared sacrifice because I would likely get fired. The reason given was that the policy not to pay them had been set by a higher level of governance.

This could be a situation faced by many OHS advisors, even at the executive level. In this case I proceeded to tell the CEO why the workforce was "rebelling" (their term) and decided to concentrate my efforts on the plant manager. It turned out well because he was open to coaching on the language he used. The most important change was that he agreed to take an empathetic approach and say that he was sorry for what they were experiencing. This was very important because he could not make his superiors lose face by saying that this wasn't his decision or that he did not agree. The employees began to open up to him and later he was able to persuade the CEO to allow him to pay the workers to come in and work on resolving safety issues that had been open for quite some time. This solution seemed to fit within corporate guidelines.

Lesson learned: While OHS advisors cannot single-handedly change organizational systems, they have the influence and power to improve the lives of those around them by providing what people most desire: to be seen and heard. An OHS advisor commented:

> These strategies are powerful and I relate to them. As a safety leader in a mining industry where the culture is deeply rooted and no one wants to change, plus I

do not have any authority. I rely on relationships and control what I can, which is getting to know the workers and strengthening my relationship with them.

(OHS Advisor)

A Steward of Organizational Memory and Learning

Making OHS a learning organization represents a shift away from the traditional, individualistic approach to incident investigation and toward a collective, systems-based model of learning and adaptation. This skill is not merely about documenting past events; it is about actively ensuring that the organization, as a living system, develops an immunization memory to fight future threats more effectively. As detailed in Chapter 10 case study, the organization possessed the "antibodies"—a blowout prevention fix—but failed to distribute them to the crew. Neither had they implemented what they learned from the BP Texas City Refinery explosion (CSB, 2016).

Shifting from Blame to Learning

One role of the Navigator is to transform incident investigations from a search for a single "root cause" and individual blame into a deep exploration of social and systemic factors. i A steward of organizational memory applies a social constructionist lens to investigations, seeking to understand:

- How people made sense of the situation at the time, given the pressures, information, and cultural norms they were operating under.
- The complex interplay of social dynamics, communication patterns, and power structures that contributed to the event.
- The shared understandings and unwritten rules that influenced risk perception and decision-making.

By asking different questions, the steward moves beyond finding a scapegoat to uncovering the systemic vulnerabilities that allowed the incident to occur, paving the way for more effective, systemic improvements. Pupulidy argues that these questions should begin by asking, "Why did it make sense for people to do what they did?" This question primes the investigation to look for influencing, or performance shaping factors that would traditionally go un-noticed (Pupulidy et al., 2023).

The "Failure to Learn" at Deepwater Horizon

The core of the "Risk Navigator" role is to see the unseen risks in the social landscape and to understand the presence of influences that may be inconsistent with protecting trust and open communication . The Risk Navigator must be the steward who prevents the 'failure to learn' seen at Deepwater Horizon, where the technical fix was known but the social pathways were blocked..

While the scale of the Deepwater Horizon disaster is beyond an advisor's direct control, the subsequent opportunity to make sense of the event from a social and systemic perspective offers an opportunity for greater learning and is perfectly within their scope of work. It illustrates why the skills of sensemaking, social construction, NTS, and stewardship of memory are not abstract ideals, but vital, life-saving competencies. It points the OHS advisors to recognize the seeds of such a failure in their own environment and navigate their organization away from them (Weick, 2001).

As Professor James Reason (1997: 223) wrote, "You cannot change the human condition, but you can change the conditions under which humans work." A cautionary note would be, "Only if you include attention to the conditions that shape human reaction and decision-making."

Making the Case for Soft Infrastructure

OHS professionals must be able to articulate the strategic value of this human-centered approach in the language of executives. Firms like McKinsey and Deloitte, provide data for executives that demonstrate that investing in trust, psychological safety, and open communication are direct drivers of business continuity, performance, and resilience, safer workplaces can emerge (Fisher & Silvergate, 2022).

However, the earlier discussion on generative leadership (Chapter 9) provides motivation when statistical data may not be sufficient. Imagine the modern safety professional's role within the immunization network is that of a bridge whose responsibility is to nurture the social health of the organization. Where social health is equal to or perhaps even greater than simply ensuring compliance with organizational controls.

The concepts in this book provide a shield; a professional truth that eliminates the need to **J**ustify, **A**rgue, Defend, **E**xplain your value. Historical research on significant industrial failures provides the gravitas to make *the intangible visible alongside the physical influences*.

Conclusion

Are you someone who is committed to a human-centered safety management approach?

Changing the way people have been doing the work is a difficult task. This isn't a task for everyone. At first facilitating these changes might feel lonely. You might feel that others are not understanding concepts that appear very simple to you. Don't get frustrated and lose patience. Succeeding in this effort takes a leader. That means that regardless of what you see going on around you remain committed to believing in people and the principles of RCS. When you model these principles, you provide people with small wins so that they can see that these principles work. Many leaders have embarked on this journey and succeeded. However it does take determination and the willingness to travel alone even if it is only at the beginning.

To break through the status quo, OHS advocates must develop skills and acquire knowledge beyond the traditional focus on physical hazards and individual behavior. By understanding how social dynamics influence risk perception and behavior, OHS advisors can drive change toward creating a workplace where people feel valued, heard, and able to contribute to organizational well-being. The goal extends beyond preventing injuries. It is a holistic approach to safety that prioritizes and recognizes the interconnectedness of physical, mental, and social health, acknowledging that each dimension can negatively impact the others and overall well-being.

The Risk Navigator metaphor effectively highlights the complexities of the OHS advisor role and the need for a broader skillset that encompasses social and psychological understanding. It offers valuable perspectives and practical guidance for individuals working in or entering the safety and health field.

Advisor Takeaways

- Redefine Your Role as a Risk Navigator. Move beyond seeing yourself as a compliance officer. Your true value lies in navigating the complex social landscape of your organization, sensing the unseen threats, and building the relationships that make your team resilient.
- Cultivate Your "Sixth Sense." Just like an ancient mariner, train yourself to be receptive to the intangible signals in the workplace—the shifts in morale, unspoken anxieties, and workarounds. Use tools like the Ladder of Inference to sharpen your awareness of your own biases and assumptions, allowing you to listen non-judgmentally and with curiosity.
- Prioritize Conversation and Connection. Understand that an organization is a conversation. Your most important tool is your ability to listen and engage in authentic dialogue. Prioritize "Ride-Alongs" and one-on-one conversations to build the trust and rapport needed to gain accurate information and empower your teams.
- Practice the Ethic of Care. Recognize that a human-centered and caring approach is your internal compass. Your ability to show empathy, build trust, and maintain dignity for everyone you interact with—regardless of their position—is the foundation of your influence and the most powerful defense against fear and silence.

Executive Takeaways

- Recognize that OHS advisors are managing complex psychosocial risks. They are your "Risk Navigators." Give them the political cover and resources to do their job.
- If you want a culture of care, you must speak it. Avoid blame-heavy language and adopt the language of collaboration.

- High attrition rates among advisors are a warning sign. If your advisors feel isolated, your immune system is failing. Ensure they are included in operations and that their psychological safety is protected.

Investing in OHS advisors and actively shaping an organizational environment where they can thrive as risk navigators is not merely a support function; it is a strategic imperative. By acknowledging the complex realities they face and providing the necessary tools, training, and leadership support, executives can significantly enhance safety outcomes, cultivate trust, and build a more resilient, equitable, and ultimately more productive workplace capable of effectively preventing SIFs.

References

Carrillo, R. A. (2024). *OHS Voices from the Resistance*. Carrillo and Associates.

Carrillo, R. A. (2025). Personal Conversation with OHS Advisor.

CSB. (2016). CSB Investigation: At the Time of 2010 Gulf Blowout, Transocean, BP, Industry Associations, and Government Offshore Regulators Had Not Effectively Learned Critical Lessons from 2005 BP Refinery Explosion in Implementing Safety Performance Indicators. U.S. Chemical Safety and Hazard Investigation Board. www.csb.gov/csb-investigation-at-the-time-of-2010-gulf-blowout-transocean-bp-industry-associations-and-government-offshore-regulators-had-not-effectively-learned-critical-lessons-from-2005-bp-refinery-explosion-in-implementing-safety-performance-indicators/

Fisher, J., & Silvergate, P. H. (2022). *The C-Suite's Role in Well-Being*. Deloitte Insights. www.deloitte.com/us/en/insights/topics/leadership/employee-wellness-in-the-corporate-workplace.html.

Goffman, E. (1959). *The Presentation of Self in Everyday Life*. Anchor Books.

INSHPO. (2017). *The Occupational Safety and Health Professional Capability Framework: A Global Framework for Practice*. www.inshpo.org/storage/app/media/docs/INSHPO_2017_Capability_Framework_Final

Pupulidy, I., & Vesel, C. (2023). *Human & Organization Potential*. Dynamic Inquiry LLC.

Provan, D., Dekker, S., & Rae, D. (2018). Benefactor or burden: Exploring the professional identity of safety professionals. *Journal of Safety Research*, 66: 21–32. https://forgeworks.com/wp-content/uploads/2022/09/Safety-Professional-Identity.pdf

Reason, J. (1997). Managing the Risks of Organizational Accidents. Ashgate.

Tetteh, C. (2025). *Organizational Learning through Storytelling*. Department of Engineering, University of Alabama Birmingham. ASEM 620: Capstone Project Part 2.

Weick, K. E. (2001). *Making Sense of the Organization*. Blackwell.

14 Conclusion

Introduction

As we reach the end of this journey, we must confront the final test of any methodology: its results. If we evaluate the current state of safety management purely by the fruit of its labor, the verdict is unsettling. Despite decades of regulation and investment, serious injury and fatality rates have stagnated. This plateau is the undeniable evidence that the "myth of individualism"—the ideology that has dominated organizational safety and health—has exhausted its utility.

The weakness of this ideology is laid bare by the numbers. However, the case studies presented in this book offer a counter-narrative. They demonstrate that a different approach, one based on Relationship-Centered Safety principles, produces vastly superior results—going past the numbers to the well-being and prosperity of the organization. Good results prove the quality of the process.

Diagnosing the Decay

This is not about a minor mistake or a policy error; it is a profound, systemic decay that must be addressed at its root. The roots, in this case, are found in the way people are treated in the workplace. The myth allowed us to ignore the social reality of work, creating a culture where liability is shifted rather than shared.

The Executive Challenge

To clear out this decay requires more than new metrics, processes, or laws; it requires a fundamental shift in executive consciousness. We must ask: Do executives truly believe that their workers are willing and able to contribute to the success of the organization? Do they understand that accountability and true communication can only manifest in an environment of inclusion and trust?

Until leadership accepts that safety is a collective outcome, created by an intricate web of human interactions, the rot will remain. We must move beyond the illusion that we can control outcomes by controlling individuals.

DOI: 10.4324/9781003668770-14

The Myth Extends beyond Safety

Safety is not unique in its obsession with the individual. In safety, we see a crash and ask, "Which individual made a mistake?"—ignoring the system. In business, we see a profit record and ask, "Which individual is the genius?"—ignoring the system. It is the same methodological flaw, just with different consequences (Blame vs. Bonus).

This flaw is supported by what James Meindl calls "The Romance of Leadership." Meindl demonstrated that the more complex and uncontrollable an organization becomes, the more people attribute outcomes to the leader. We credit (or blame) the individual because we cannot understand the system. Humans have a psychological need to feel that events are controllable. Admitting that a system (or an accident) is complex and random is terrifying. Blaming an individual or praising a hero restores our sense of order.

The Overarching Purpose

The overarching purpose of this book is to provide a roadmap for Relationship Centered Safety—moving it from a mechanistic, compliance-driven function to a robust, human-centered social science. This evolution is mandated by the persistent failure of individualism, an ideology confirmed globally to deflect corporate leadership's governance accountability for systemic organizational failures onto the individual at the sharp end.

The tragedy of the Deepwater Horizon (DWH) offers a stark, final example of the destructive nature of this myth. On the day of the disaster, the BP Vice President of Drilling, Patrick O'Bryan, was on the rig for a "VIP Visit" but was never held accountable for the systemic failures designed in the corporate office. Instead, the Department of Justice's prosecution sought to translate vast organizational failures into the "criminal gross negligence of two individuals": well site leaders Don Vidrine and Robert Kaluza.

Kaluza, who was prosecuted despite the executive failures, argues in his book *Deepwater Deception* that the entire legal process was an act of "legal theater" and a "scapegoat mechanism." His prosecution attempted to divert public attention from the senior executives responsible for overseeing the design of the flawed systems, cost-cutting measures, and ambiguous safety protocols that were the root causes of the disaster. This legal strategy—holding the individuals on the rig criminally liable while the executives who ratified the flawed system remained unscathed—perfectly illustrates the danger of an individualistic worldview. It is a convenient fiction that allows the system to remain unchanged.

For too long, we have relied on this myth that safety is primarily a matter of individual accountability. We have exported this "Made in the USA" philosophy to every corner of the industrial world, creating a global standard that weakens risk management. It contributes to a fear-driven culture that stifles the open communication essential for preventing catastrophe.

The Cost of Fragmented Awareness

The ultimate consequence of maintaining a culture rooted in the myth of individualism is the crippling of the organization's capacity for collective sensemaking. The DWH disaster was not a failure of an isolated malfunction but a systemic failure—a complete breakdown in trust and communication.

This disrupted the organization's immune system that transmits the embodied knowledge and emotions necessary for situational awareness. A prime example was the failure of risk management at the executive level. After it was too late, investigations discovered that a blowout prevention fix had been shared by Transocean with BP corporate but was never communicated to the DWH crew.

If the failure to learn continues, the next disaster will also be a product of systemic ignorance. Future crises or catastrophes will not just be accidents but direct outcomes of widespread, deeply ingrained failures to recognize or act upon systemic vulnerabilities. This represents the antithesis of stewardship. As long as organizations operate under the ideology of individualism, they are structurally designed to silo life-saving information. The belief that "I am responsible" justifies compartmentalization, guaranteeing that the failure to learn remains perpetual.

The Occupational Health and Safety (OHS) Advisor Crisis

The OHS advisor's lived experience is one of profound conflict. While 67% of their job consists of relational, social, and psychological activities—they often lack the professional mandate or resources to execute this role effectively. This disparity leads to high rates of burnout, exclusion, and resignation, leading to 55% of OHS advisors thinking of quitting in 2024.

High attrition among OHS professionals is not merely an HR problem; it is a warning sign that the organizational immune system is failing. When OHS advisors resign due to emotional labor or feeling unheard, the organization loses its "immunization memory." The role of OHS advisor– redefined as Risk Navigator—fits professionals who navigate ongoing unpredictable conflicts, emotions, fears, and power plays. Their greatest influence is achieved by prioritizing communication and providing what people most desire: to be seen and heard.

Leader as Social Architect

The reality is that safety is a collective outcome, created by an intricate web of human interactions, trust, and shared meaning. By adopting a Relationship-Centered worldview, the leader moves from being a monitor of behavior to a social architect who builds the trust and shared meaning necessary for a resilient organization."

Safety's effectiveness hinges on a fundamental truth: "True communication takes place only in the presence of relationship and trust." Consequently, the efficacy of every conventional safety tool is directly proportional to the strength of the underlying social relationships. This inescapable reliance elevates safety to a social science, requiring a leader who functions as a social architect.

This role requires a commitment to the Paradox of Leadership: practicing selfless empathy to build a relationship-based culture while maintaining the inner conviction to act decisively when required. It requires the non-technical skills to "read the room." The primary obstacle to connecting with others—and seeing risk signals—is the fear of losing face. It can be overcome by a simple question: "Tell me what is happening."

A Generative Future

We are at a crossroads. The path of least resistance—compliance, blame, and individualism—has led us to a plateau where fatality numbers refuse to drop. To break through, we must embrace a new reality: the OHS profession is evolving into a "helping profession," and safety is evolving into a social science.

Change does not happen through new rules or slogans. It happens through Radical Empathy—the discipline of setting one's own agenda aside to truly understand the other's reality. The Rio Tinto GM in our case study didn't single-handedly convert the organization's belief system; he used his authority to mandate simple behaviors that created the psychological safety needed for a new culture to emerge. That is Relationship-Centered Leadership at work.

The Agent of Change

It is time to abandon the overuse of individualism and the mechanistic approach to managing risk. Instead, choose a relationship-centered strategy and generativity. If you choose the role of change agent your influence and impact will be directly proportional to the relationships you develop to spread and enact this message.

Afterword

From Individual Responsibility to Collective Care

I have spent almost 30 years working with leaders, organizations, and safety professionals across continents, cultures, and industries. From mining camps in remote regions, to boardrooms in global corporations, to frontline operations where the consequences of decisions are felt immediately and physically, one truth has followed me everywhere: people do not get hurt at work because they do not care about safety. They get hurt when systems, relationships, and cultures make it difficult—or impossible—for care to be acted upon.

This matters, because for decades our field has been built on good intentions. The emphasis on personal responsibility emerged from a genuine desire to empower individuals, to encourage vigilance, and to prevent harm. It was never malicious. But good intentions do not guarantee good outcomes. As *Unmasking Safety's Myth of Individualism* makes uncomfortably clear, the limits of an individualistic approach to safety have now been reached.

This Is Why Rosa Carrillo's Work Is so Timely

I have known Rosa for several years as both an esteemed colleague and a trusted voice in our field. She serves on the Board of the International Institute of Leadership & Safety Culture, where I am CEO and Chairman of the Board, and her thinking consistently challenges leaders to move beyond familiar technical solutions and ask harder, more human questions. *Unmasking Safety's Myth of Individualism* is the most complete expression of that challenge.

What Rosa does here is not to dismiss personal responsibility, but to place it where it belongs: *within* a social system. She shows, with clarity and rigor, that safety is not an individual achievement but a collective outcome—shaped by power, trust, relationships, and the everyday conditions in which work is done. When we focus too narrowly on individuals, we often miss the very forces that make safe behavior possible or impossible in the first place.

I have seen this dynamic play out repeatedly around the world.

I have worked with organizations where people knew the rules, understood the risks, and genuinely wanted to do the right thing—yet still remained silent when it mattered most. Not because they were careless, but because the social cost of

DOI: 10.4324/9781003668770-15

speaking up was too high. I have listened to safety professionals describe the quiet erosion of influence that comes from being held responsible for outcomes they do not control. And I have seen leaders express real surprise when their stated commitment to safety was not reflected in what people actually did on the ground. These are not failures of competence. They are failures of connection.

One of the most important contributions of *Unmasking Safety's Myth of Individualism* is its insistence that safety is a social science. It is produced through interaction, not instruction. It emerges—or fails to emerge—through how leaders listen, how trade-offs are discussed, how mistakes are treated, and how power is exercised when pressure rises. Safety does not live in policies. It lives in conversations. This is where psychological safety becomes central.

Psychological safety is often misunderstood as a soft or permissive concept. In practice, it is neither. It is the shared belief that speaking up, asking questions, admitting uncertainty, or challenging assumptions will not result in embarrassment, punishment, or exclusion. Rosa is clear that psychological safety is not created by slogans or workshops; it is a social achievement that emerges from consistent leadership behavior over time.

For leaders, this reframes the role entirely.

The leader's task is no longer just to manage systems or enforce compliance, but to act as a social architect: to design the conditions in which people can speak honestly, raise concerns early, and learn together. Rosa's Relationship-Centered Safety framework provides leaders with a practical way to do this. It translates decades of social and organizational research into beliefs and behaviors that can be practiced every day—not abstract ideals, but operational disciplines.

What does this mean in practice?

It means paying attention to what is rewarded, not just what is stated.

It means noticing who speaks in meetings—and who does not.

It means how incidents and near misses are discussed: whether conversations close down around blame or open up around learning.

It means whether safety advisors are treated as strategic partners or technical specialists brought in after decisions are already made.

One relationship centered belief in particular reflects what I have seen repeatedly across cultures: *people will speak up if it is in their interest to do so.* This is not a moral judgment; it is a realistic one. People constantly assess risk—physical, social, and emotional. If speaking up threatens their credibility, relationships, or livelihood, silence becomes a rational response. Policies like stop-work authority fail not because people do not understand them, but because the surrounding culture does not support their use.

When psychological safety is present, something important happens: organizations become more resilient.

Resilience is often discussed as the ability to bounce back after failure. In reality, resilient organizations do something more important—they detect weak signals early, adapt under pressure, and learn continuously. This capacity does not come from procedures alone. It comes from relationships that allow information to flow freely, especially when the message is uncomfortable.

In resilient organizations, people speak sooner, not later. Problems are surfaced while they are still manageable. Learning happens before harm occurs. Rosa's work shows clearly that resilience is not a structural feature—it is a relational one. It depends on trust, inclusion, and a shared commitment to looking after one another.

This Is Where Care Becomes Central

Care is not softness. It is not indulgence. And it is certainly not the absence of accountability. Care is the precondition for accountability that actually works. In every organization I have seen achieve sustained safety performance and resilience, there is a strong ethic of care: leaders listen before they judge, people are treated with dignity when things go wrong, and trust is built deliberately over time.

Cultures of care do not emerge by accident. They are designed—through beliefs, structures, and daily leadership behavior. Rosa's socio-technical systems iceberg makes visible what is usually ignored: the submerged layers of assumptions, relationships, and power that determine whether technical systems function or fail. Importantly, *Unmasking Safety's Myth of Individualism* does not argue for abandoning compliance or engineering excellence. It shows why those efforts only succeed when the social foundations are sound.

For OHS professionals, this book offers both validation and challenge. Validation, because it names the relational barriers many have experienced but rarely felt safe to articulate. Challenge, because it calls for a shift in role—from enforcer or technical expert to trusted partner and influencer. That shift requires non-technical skills: communication, sensemaking, emotional intelligence, and courage. These are not "soft" skills; they are essential ones.

The concept of generativity may be the most powerful invitation in this work. For leaders, especially those later in their careers, the question shifts from control to legacy. What kind of organization are you leaving behind? One where people protect themselves and stay silent—or one where they look out for one another, speak early, and recover quickly when things go wrong?

Unmasking Safety's Myth of Individualism will not give you a checklist. It will give you a way of seeing. Once you understand safety as a collective, relational outcome, it becomes very difficult to return to the simplicity of individual blame. You begin to notice the invisible forces shaping behavior. You ask better questions. And over time, you design organizations that are not only safer, but more resilient, more adaptive, and more humane.

That is what a culture of care looks like in practice.

Rosa Carrillo has written a book that is urgently needed. It does not flatter the OSH profession, and it does not offer shortcuts. Instead, it invites leaders and safety professionals to mature—to take responsibility not just for outcomes, but for the systems, relationships, and conditions that produce them.

If you accept that invitation, what becomes possible is significant: safer people, stronger teams, fewer injuries, greater resilience, deeper trust, and organizations capable of learning before harm occurs.

I believe that is a legacy worth choosing.

Dr Andrew Sharman
CEO and Chair, International Institute of Leadership & Safety Culture
Co-Author of *Mind Your Own Business: What Your MBA Should Have Taught You About Workplace Health & Safety and Who Cares Wins: The Psychological Science of Transformational Leadership*

Appendices

Introduction

We have chronicled the journey of safety from a mechanistic, compliance-driven function to a robust, human-centered social science. Chapters 1 through 14 established the compelling *why* (the flaw of individualism) and the *what* (Relationship-Centered Safety or RCS).

The core challenge for leadership remains: How do we translate this essential shift in philosophy into measurable, sustainable action?

To maintain the flow of the narrative, the comprehensive tools, detailed procedures, and critical reference materials required for this implementation phase have been organized here as a set of appendices.

Appendix A
The Risk Navigator Job Description: Chapter 13

Purpose: To redefine the occupational health and safety (OHS) role from a "Compliance Officer" to a "Risk Navigator." Use this for hiring and performance reviews.

The mind-mapping exercise revealed that 67% of the OHS advisor's responsibilities fall into the soft infrastructure category. The complete Risk Navigator Job Description, found here, translates this reality into a formal mandate for strategic hiring and performance review, moving the role beyond that of a compliance officer to a navigator of complex social and psychological risks.

Core Competency Breakdown (For HR and Hiring)

When hiring or training a Risk Navigator, look for these specific proficiencies beyond standard safety certifications:

1 Social Perception (The "Sixth Sense")

- *Definition*: The ability to detect "invisible" risks—shifts in morale, unspoken anxieties, and the "saving face" dynamics that silence crews.
- *Behavior*: Notices when a team is in "survival mode" versus "self-expression mode" (Inglehart-Welzel).

2 Polarity Management

- *Definition*: The ability to manage unsolvable tensions (e.g., Safety vs. Efficiency or Compliance vs. Autonomy) without forcing a false "either/or" choice.
- *Behavior*: Facilitates "Both/And" strategies (e.g., "We will be highly compliant and highly adaptive").

3 Strategic Communication (Embodied Language)

- *Definition*: The ability to speak the language of the workforce (lived experience) and the language of the executive (business continuity).
- *Behavior*: Translates "soft infrastructure" needs into hard business cases (return on investment, retention, resilience).

Table A.1 Risk Navigator job description

Domain of practice	*The traditional view (compliance officer)*	*The Risk Navigator view (social architect)*
Primary Mandate	Enforce Compliance: Ensure adherence to rules, regulations, and procedures to avoid penalties.	Nurture Resilience: Monitor and strengthen the organization's "immune system" (relationships and trust) to navigate uncertainty.
Scope of Risk	Physical and Tangible: Focuses on slips, trips, machinery, chemicals, and regulatory violations.	Socio-Technical and Invisible: Includes physical risks plus psychosocial hazards, power dynamics, fear of losing face, and production pressures.
Relationship to Workforce	The Police/ Enforcer: "I am here to catch you doing something wrong." Interactions are transactional (audits).	The Partner/Facilitator: "I am here to help you navigate complexity." Interactions are relational (Ride Alongs).
Key Activity	Inspections and Audits: Checking boxes against a static standard.	Sensing and Sensemaking: Detecting "weak signals" (low-grade fevers) in the culture and facilitating dialogue to understand them.
Response to Error	Find the "Root Cause": Often stops at "Human Error" or "Failure to follow procedure."	Understand the Context: Asks, "Why did this make sense to them at the time?" Looks for systemic traps.
Core Competencies	Technical Knowledge: Regulations, industrial hygiene, engineering controls.	NTS: Social perception, emotional intelligence, polarity management, and linguistic framing.
Measure of Success	Lagging Indicators: TRIF, LTI rates, audit scores.	Leading Indicators: Trust levels, frequency of SWA ("I've got a question"), quality of collaboration.
Communication Style	Over-Explaining (JADE):Tries to "convince" or "soften" the message toensure the manager understands and isn't upset.	Boundary-Setting: States the professional truth clearly and objectively, accepting that the reaction belongs to the receiver.
Reaction to Resistance	Internalized Failure: "I didn't explain it well enough," leading to emotional exhaustion and repeated attempts to "fix" the listener.	Radical Acceptance: "I have provided the risk assessment. The decision—and the liability—now rests with you."
Source of Self-Worth	External Validation: Worth is tied to whether the organization accepts the advice and achieves "Zero Harm".	Internal Integrity (Wholeness): Worth is tied to remaining a "safe" person for the crew and speaking the truth without distortion.
Role in Crisis	The Savior/Fixer: Trying to control the outcome and carrying the anxiety of the entire organization.	The Calm Center: Focusing on remediation and support without taking on "unearned guilt" for systemic failures.

4 Generativity (The Ethic of Care)

- *Definition*: A drive to guide and support the well-being of the collective, rather than just enforcing rules.
- *Behavior*: Prioritizes "Inclusion precedes Accountability" (RCS Belief #2).

Facilitator's Note for Leaders

When using this tool during performance reviews, do not throw out the "Traditional" column entirely. Compliance is still the floor—but the Risk Navigator builds the ceiling. Ask your safety team: "What percentage of your week is spent auditing (Left Column) vs. sensing and relating (Right Column)? How can we shift that balance to 50/50?"

Appendix B

Closing the Linguistic Divide

The Strategic Language Phrasebook

Purpose: To shift leadership language from "Systemic/Abstract" to "Embodied/Human."

Usage: Use to prepare for town halls, difficult 1:1s, or email communications.

Review the following examples and identify questions from surveys or town halls that keep coming up.

Scenario 1: Addressing Burnout and Production Pressure

The compliance response (Abstract)	*Why it fails*	*The embodied response (human-centric)*	*Why it works*
"We understand concerns that production goals overshadow well-being. We are assessing capacity constraints."	Why it fails:	"I hear your frustration. I know you are tired, and I know the current pace feels unsustainable. We are going to build a workload plan together, because your physical well-being is the foundation of our success."	Why it works: Validates the physical/emotional reality.

Scenario 2: Responding to "people are afraid to raise questions"

The compliance response (Abstract)	*Why it fails*	*Embodied response (human-centric)*	*Why it works*
"We have a genuine desire to hear from you. The leadership team continues to explore multiple avenues to facilitate open and honest dialogue	Corporate jargon ("avenues," "facilitate").	We recognize the negative potential of a chilled work environment and are actively working with our supervisors and managers to improve relationships by listening to and responding to concerns.	Uses sensory language ("chilled" "listening" "responding").

Appendix C

The NTS Leadership Audit

Purpose: To self-assess the critical "Non-Technical Skills" (NTS) required for high-hazard leadership.

Instructions: Leaders rate themselves (1–5) on the following competencies.

1 Social Perception (The "Sixth Sense")

[] I can tell when the room has gone silent due to fear rather than agreement.
[] I notice changes in body language when production pressure increases.
[] I can pick up on the general emotional tone of a meeting or the office, helping them tailor their approach.

2 Saving Face (Dignity Management)

[] I deliver corrective feedback in private to prevent humiliation.
[] I create opportunities for people to admit mistakes without looking incompetent.

3 Collective Sensemaking: Good sensemaking involves a set of cognitive and social skills and techniques used to understand ambiguous or complex situations and make informed decisions.

[] When a problem arises, I ask: "What do you know that I need to know?"
[] I avoid imposing my answer first; I let the team construct the solution.
[] I notice people who aren't speaking and ask them to speak.
[] I draw inferences and check them with the team.

4 Situational Awareness

[] I track not just the *task*, but the *stress level* of the team performing the task.
[] I recognize when "business as usual" has shifted to "survival mode."
[] I analyze the information I perceive to understand the current situation and its meaning.
[] I practice staying present and fully engaged with my surroundings.
[] I pay close attention to what others are saying to pick up on important details or warnings.

Appendix D

The "Ladder of Inference" Workshop

Purpose: To become aware of how making assumptions prevents understanding others and can lead to bad decisions.
Format: 60-minute team discussion.

The Facilitator's Script: Review the Ladder of Inference.

"We often assume our beliefs are obvious facts. But usually, we are filtering reality through our own experiences. Today, we are going to look at a recent breakdown to trace how two groups climbed the 'Ladder of Inference' to reach opposite conclusions."

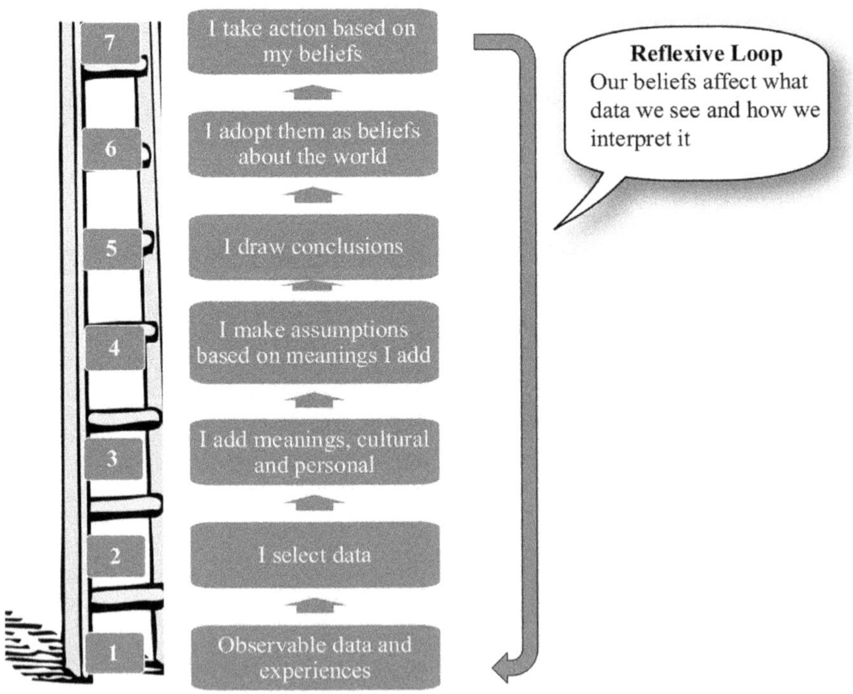

Case Study: The Strike (Visual Guide)

Distribute Table D.1 to the team to illustrate how one event creates two realities.

Table D.1 Example of ladder dynamics

Rung of the ladder	*The plant manager's reality* *I see*	*The workforce's reality* *We see*
1 Observation	Manager announces no repercussions for reporting near misses	We report a near miss.
2. Select Data	Saw a violation of lock out tag out.	Saw a disciplinary letter in a personnel file.
3. Draw Conclusion	"The 'no punishment' promise cannot apply to life-threatening violations."	"Management broke their promise; they cannot be trusted."
4. Adopt Belief	"I must hold people accountable to save lives."	"We must retaliate to protect ourselves from management power."
5. Take Action	Issues a disciplinary letter.	We call a strike.

The Team Challenge

Ask the group to identify a current friction point. Draw the ladder and identify what goes on each rung. Ask: "What data are 'they' seeing that 'we' might be missing?"

Appendix E

Social Construction Audit: Chapter 8

Purpose: To identify the invisible rules and power dynamics shaping your team culture.

Format: Leadership Team Offsite Exercise.

Table E.1 Principles of social constructions

Social principle	*Diagnostic question to ask your team*
Power Dynamics	Who is allowed to speak first in our meetings? Who is interrupted? Does the "expert" or the "boss" define the risk?
Multiple Realities	Do we have a "Head Office" reality vs. a "Frontline" reality? When was the last time we validated our view with a *Ride Along* or *Whistle Stop Tour*?
History and Culture	What is the "legend" of our safety culture? Do we tell stories of people who stopped work (heroes) or people who met quotas (heroes)?
Shared Meaning	When we say "Accountability," do employees hear "Ownership" or do they hear "Blame"?
Relational Foundation (RCS Belief #1)	How is trust actively built or damaged in our daily, one-on-one interactions? (e.g., Are commitments kept? Is an "attentive gaze" used?)
Psychological Safety and Voice	Where in our processes is it most difficult for an employee to ask a question or stop a job without feeling embarrassed or disrespectful? (The "Saving Face" Audit)
Inclusion and Accountability (RCS Belief #2)	How do we actively demonstrate that Inclusion precedes accountability? Do we ensure people feel seen and heard before expecting them to take responsibility for systemic issues?

Appendix F

Evaluating and Improving Communication in Social Interactions

Purpose: Provides a diagnostic tool for observing and evaluating five types of interactions contributing to unwanted events (e.g., Team Communication, Power Dynamics, Quality of Relationships)

Table F.1 offers a framework to understand and evaluate the interactions between individuals within an organization or any social system. They move beyond simply observing what people do and delve into the why and how of their actions within a complex web of influences. It also questions how information is interpreted by the listener, to better understand the origin of miscommunications.

Evaluating five types of interaction is a tool for discerning how communication breakdowns, power dynamics, and the absence of psychological safety may have contributed to unwanted events. Bea's (2021) emphasis on the "failure to learn" directly relates to the significance of conversation as a skill and the exploration of organizational barriers to dialogue. Had these organizations invested more in communication skills and greater awareness of diverse perspectives, they might have more effectively identified and addressed escalating risks.

Table F.1 Evaluating interactions

Type of interaction	*Evaluation questions*	*Application example*
1. Team Communication and Information Flow	• How is information shared (or not shared)? Observe the formal and informal communication channels. • Are there bottlenecks? Is information transparent and accessible, or is it siloed? • Who has access to what information, and how does this affect their understanding and actions? • Look for instances where lack of information or miscommunication might have influenced a decision or behavior.	In a safety incident review, instead of just focusing on the individual's actions, you could investigate if they had the necessary information about potential hazards or the correct procedures. Was this information readily available and clearly communicated? Did the individual understand the information correctly?

Table F.1 (Continued)

Type of interaction	*Evaluation questions*	*Application example*
2. Influence of Social Norms and Expectations	• Pay attention to the unwritten rules of behavior. What is considered acceptable or unacceptable within the team or organization? • Are there subtle pressures to conform to certain ways of doing things, even if they aren't formally stated? • Observe how newcomers learn "how things are really done."	In a team setting, notice if there's an unspoken norm to complete tasks quickly, even if it compromises quality or safety. This might explain why an individual took a shortcut, even if they knew it wasn't the best practice. • As a team exercise, have people answer these questions anonymously and compare answers. This will make for a great discussion.
3. Power Dynamics	• Analyze the formal and informal hierarchies and power structures. How do these dynamics influence communication, decision-making, and the willingness of individuals to speak up or challenge the status quo? • Are certain individuals or groups more influential than others?	Observe if junior team members are hesitant to voice concerns or disagree with senior members, even if they have valid points about a potential risk. This power dynamic could stifle crucial feedback. • Arrange for a dialogue between groups where power differences interfere with trust or open communication. Use questions on the left.
4. Group sensemaking	• How do individuals collectively interpret events, risks, and problems? • Are there shared narratives or assumptions that might be influencing their actions? • Observe group discussions and how shared understandings are constructed.	If a team has a shared belief that certain risks are unavoidable, this collective sensemaking might lead individuals to be less proactive in identifying and mitigating those risks. • Offer an anonymous way to contribute. A lack of psychological safety might lead to a superficial analysis that doesn't uncover the root causes of issues.
5. Quality of relationships and the presence of psychological safety	• Observe the level of trust, respect, and openness among team members. • Do individuals voice concerns, admit mistakes, ask questions, or challenge ideas without fear of negative repercussions? • High-quality relationships and psychological safety foster better communication and learning.	In a post-project review, assess if team members felt comfortable sharing what went wrong and offering constructive criticism. Offer an anonymous way to contribute. A lack of psychological safety might lead to a superficial analysis that doesn't uncover the root causes of issues.

Appendix G

Social Construction of Risk Principles Applications

Purpose: Translates six principles of social construction (e.g., Understanding of Risk, Power Dynamics, Multiple Realities) into specific, actionable leadership actions for application in the workplace.

The principles of social construction outline the six principles of social construction and their application in the workplace. The first column is the principle. The second is what it means in the context of social construction. The third is how it looks in application. The leadership action row after each principle shares specific leadership activities to enact the principle.

Applications of Six Principles of Social Construction Worksheet

Principle	*Meaning*	*Application*
1. Understanding of risk	Risk is subjectively perceived and interpreted by individuals based on their experiences, social interactions, and cultural context.	Actively seek to understand how workers perceive and make sense of risk in their specific work environment. Checklists and procedures are only as good as the last time there was a discussion about what they mean.
Leadership action		Include frontline workers in the risk assessment process, allowing them to share their knowledge and perspectives on potential hazards and solutions.

Principle	*Meaning*	*Application*
2. Communication and shared meaning- making	Clear and effective communication is crucial for creating a shared understanding of risk.	Use clear and inclusive language: Avoid jargon and technical terms. Frame safety messages in a way that resonates with workers' experiences and values. Articulate the larger purpose and how people fit in it.
Leadership action		Encourage conversations and open dialogue: Create a safe space for workers to voice their concerns, ask questions, and challenge existing practices without fear of reprisal on a regular basis.
3. Multiple realities and situational awareness	There is no single, objective reality. Individuals and groups may interpret the same situation differently. Knowing what is going on, so you can figure out what to do next.	Situation Awareness, ask three key questions: What do they know that I need to know? What do I know that they need to know? What do none of us know that we need to know?
Leadership action		Encourage perspective-sharing: Create opportunities for workers such as a Q&A website to share their perspectives and interpretations of safety-related issues in real time.
4. Power dynamics	Power dynamics influence how risk is perceived and managed.	Implement policies and practices that value diversity, equity, and inclusion.
Leadership action		Give workers a voice in safety-related decisions. Encourage them to report hazards, suggest improvements, and participate in safety committees.

Principle	*Meaning*	*Application*
5. Knowledge and action	Our understanding of risk shapes our actions, and our actions shape our understanding.	Provide opportunities for continuous learning: Offer training, workshops, and mentorship programs to keep workers' knowledge and skills up to date.
Leadership action		To surface drift, create a safe space for workers to experiment with new ideas and approaches to safety. Recognize workers and incorporate new techniques.
6. History and culture	The historical and cultural context in which safety practices are embedded influences how risk is perceived and managed.	The national culture influences how people respond to a safety approach. A culture that has high power distance is more likely to accept top-down directives. A culture low in power distance resists top-down decisions and may expect engagement.
Leadership action		Implement programs that actively encourage and reward creative safety solutions, and tailor communication strategies to respect the cultural backgrounds of the workforce.

Appendix H

Glossary of Terms

The glossary serves to bridge the "linguistic divide" and ensure the specialized socio psychological and organizational development vocabulary is fully understood, particularly by those accustomed to traditional, technical safety language.

Terminology	*Definition*	*Supporting Source*
Embodied Cognition	The theory that meaning and understanding are fundamentally rooted in our bodily, physical, and emotional experiences with the environment, explaining why frontline workers' language is often tangible and emotional.	Chapter 7
Generativity	The deep, intrinsic motive (the "powerful why") to support, guide, and benefit future generations. This motive fuels the ethic of care and provides the moral basis for prioritizing people over profit.	Chapter 9
NTS	Cognitive and social skills (e.g., situational awareness, teamwork, decision-making, social perception) that complement technical skills to improve safety and operational efficiency in high-stakes environments.	Chapter 4
Path of Least Resistance	A systemic dynamic where conformity, organizational pressures, or the avoidance of social risk (like "losing face") guides behavior toward unsafe practices because it appears simpler and less risky (socially or politically) than actively challenging the system. It is produced by the system's design, not individual choice.	Chapter 1

Terminology	*Definition*	*Supporting Source*
Psychological Safety	A group dynamic or shared belief among team members that the environment is safe for interpersonal risk-taking, such as speaking up, asking questions, or admitting mistakes, without fear of embarrassment or negative repercussions.	Chapters 1, 2
Polarity Management	The ability to manage unsolvable tensions (e.g., Safety vs. Efficiency or Compliance vs. Autonomy) without forcing a false "either/or" choice.	Chapter 4
Social Relations of Work	The sociological framework asserts that the cause of accidents is not individual error, but the organization's structural design, including systems of rewards, command, and organization.	Chapter 1
Social Construction	The process by which reality, including concepts like safety and risk, is collectively created and maintained through human interaction, shared meaning, language, and cultural norms.	Chapter 8
Sensemaking	The collective and continuous process by which individuals and groups use NTS to construct shared meaning from ambiguous and complex experiences, which is essential for survival and managing the unexpected.	Chapter 8
Saving Face	The universal human need to maintain dignity, respect, or a positive social self-image within a group. The fear of losing face is the primary social barrier that neutralizes Stop Work Authority (SWA) and silences employees.	Chapter 6
Soft Infrastructure	The core strategic asset of an organization is defined as the vital, often intangible network of relationships, communication pathways, shared knowledge, and norms that determine organizational health and resilience.	Chapter 9
Social Perception	A critical NTS defined as the ability to understand the emotions and intentions of others. It is emphasized as the strongest predictor of a senior manager's effectiveness and influence on safety, not technical knowledge.	Chapters 8, 10

Index

Note: Figures are indicated by *italics*. Tables are indicated by **bold**. Endnotes are indicated by the page number followed by "n" and the note number e.g., 111n5 refers to note 5 on page 111.

For Product Safety Concerns and Information please contact our EU representative GPSR@taylorandfrancis.com
Taylor & Francis Verlag GmbH, Kaufingerstraße 24, 80331 München, Germany

www.ingramcontent.com/pod-product-compliance
Ingram Content Group UK Ltd.
Pitfield, Milton Keynes, MK11 3LW, UK
UKHW020006200926
475702UK00015B/2510

* 9 7 8 1 0 4 1 1 3 2 3 6 3 *